# The Theory of Generalized Inverses Over Commutative Rings

## Algebra, Logic and Applications

A series edited by

R. Göbel

*Universität Gesamthochschule, Essen, Germany*

A. Macintyre

*University of Edinburgh, UK*

**Volume 1**
Linear Algebra and Geometry
*A.I. Kostrikin and Yu I. Manin*

**Volume 2**
Model Theoretic Algebra: With Particular Emphasis on Fields, Rings, Modules
*Christian U. Jensen and Helmut Lenzing*

**Volume 3**
Foundations of Module and Ring Theory: A Handbook for Study and Research
*Robert Wisbauer*

**Volume 4**
Linear Representations of Partially Ordered Sets and Vector Space Categories
*Daniel Simson*

**Volume 5**
Semantics of Programming Languages and Model Theory
*M. Droste and Y. Gurevich*

**Volume 6**
Exercises in Algebra: A Collection of Exercises in Algebra, Linear Algebra and Geometry
*Edited by A.I. Kostrikin*

**Volume 7**
Bilinear Algebra: An Introduction to the Algebraic Theory of Quadratic Forms
*Kazimierz Szymiczek*

**Please see the back of this book for other titles in the Algebra, Logic and Applications series.**

# The Theory of Generalized Inverses Over Commutative Rings

K.P.S. Bhaskara Rao

*Southwestern College, Winfield, Kansas, USA*

London and New York

First published 2002 by Taylor & Francis
11 New Fetter Lane, London EC4P 4EE

Simultaneously published in the USA and Canada
by Taylor & Francis Inc.,
29 West 35th Street, New York, NY 10001

*Taylor & Francis is an imprint of the Taylor & Francis Group*



*Publisher's Note*
This book has been prepared from camera-ready copy supplied
by the author
Printed and bound in Great Britain by Biddles Ltd, Guildford and King's Lynn



*British Library Cataloguing in Publication Data*
A catalogue record for this book is available from the British Library

*Library of Congress Cataloguing in Publication Data*
A catalogue record has been requested

ISBN 0–415–27248–3

*for*
*two great kids*
*my son, Swastik*
*and*
*my daughter, Swara*

# Contents

# Foreword

A possible definition of a "generalized inverse" of a linear operator $A$ is an operator that has some useful inverse properties, and reduces to the inverse of $A$ if $A$ is invertible. The "useful properties" include solving linear equations $A\underline{x} = \underline{a}$, selecting a particular (e.g. least norm) solution if the equation has more than one, or producing an approximate solution (e.g. a least squares solution) if the equation does not have any.

Generalized inverses first arose in analysis in the study of integral equations (the "pseudo-inverse" of I. Fredholm 1903, and the "pseudo-resolvent" of W.A. Hurwitz, 1912). Although determinants, and limits of determinants, were used in some of these early studies (with linear operators being approximated by infinite matrices), the algebraic nature of generalized inverses was established later, in the works of E.H. Moore (1912, 1920, 1935), R. Penrose (1955), Drazin (1958) and others. Thousands of articles on generalized inverses appeared since Penrose's seminal article, and most of them deal with matrices over the real and complex fields. This sufficed for many applications, and required only tools from linear algebra.

Several researchers (notably D.R. Batigne, F.J. Hall, I.J. Katz, D.W. Robinson, R. Puystjens, R.B. Bapat, K.M. Prasad and K.P.S. Bhaskara Rao) studied generalized inverses in more general algebraic settings, fields and rings. The ever growing importance of discrete mathematics in modern applications have made their results timely.

Professor K.P.S. Bhaskara Rao has collected the above results, until now scattered in the research literature, and added new ones, in this concise and well-written research monograph. I expect it to advance our knowledge of generalized inverses over fields and rings, by promoting and guiding future research.

Adi Ben-Israel<br>
*RUTCOR, Rutgers University*<br>
*USA*

# Preface

My interest in the theory of generalized inverses was formed in my student days by my teacher Professor C.R. Rao. Since then I have learnt about many of the interesting features of $g$-inverses from the monographs of Ben-Israel and Greville and of Rao and Mitra. I was always more interested in the $g$-inverses of matrices over various algebraic structures than over classical real or complex fields.

The present monograph is the result of my endeavor to present to the mathematical community various aspects of the theory of $g$-inverses of matrices over commutative rings. Though this subject is relatively young, it has many beautiful results and its development has reached a final and complete stage.

The theory of generalized inverses of real or complex matrices is a well developed subject and the results of this theory have been chronicled in several monographs. See, for example, *Generalized inverses: theory and applications* by Adi Ben-Israel and Thomas N.E. Greville, Wiley (1974), *Generalized inverse of matrices and its applications* by C.R. Rao and S.K. Mitra, Wiley (1971) and the compendium *Generalized inverses and applications* edited by M.Z. Nashed, Academic Press (1976).

In algebra, though the concept of regularity (that an element $a$ in a ring $R$ is regular if there is a $g$ in $R$ such that $aga = a$) was not new, it was studied very little with respect to matrices until about twenty years ago. In the mid thirties, Von Neumann proved that if every element of an associative ring $R$ is regular then every matrix over $R$ is regular. The problem of characterizing regular matrices over commutative rings was raised by the author in [20]. This became all the more important because of the interest of control theorists and systems theorists in polynomial matrices (see [8], [52], [93] and [94] and the references therein) and that of mathematicians working in operator algebras (see [43], [44], [46], [47] and [54] and the references therein).

It was in the early eighties that problems were raised as to how much of the theory of $g$-inverses can be developed over the ring of integers; and the pursuit continued to principal ideal domains, integral domains and general commutative rings. I have presented the development as it happened giving the reader an insight into the intricacies of the subject.

Mathematicians working in $g$-inverses of matrices, algebraists, system theorists and control theorists would be interested in the results presented here. Economists also deal with polynomial matrices and this monograph should be useful for them too. The results given here for matrices over Banach algebras would be of interest to mathematicians working in operator theory.

This monograph can be used to present a one or two semester course on $g$-inverses for final year undergraduate students who are interested in algebra. It can also form the basis for a sequel to algebra and linear algebra courses. There are several monographs on matrices over rings: *Integral matrices* by Morris Newman, Academic Press (1972). *Linear algebra over commutative rings* by B.R. McDonald, Marcel Dekker (1984), and *Matrices over commutative rings* by W.C. Brown, Marcel Dekker (1993). Besides being of independent interest the present monograph acts as a sequel to these monographs.

Graduate students can utilize this to explore the yet unsolved problems (similar) for matrices over general associative rings.

Several exercises, some of which are taken from the literature, are devised to enhance the understanding of the subject. A novel feature of this monograph is an annotated bibliography which also serves as "notes and comments" to the text.

Dr. K.M. Prasad, many of whose results are also presented here, helped me in the initial stages of the preparation of this monograph.

Professor Adi Ben-Israel and Professor D.W. Robinson kindly looked at the manuscript and I thank them for their constructive criticism and advice.

Most of the work on this monograph was done while the author was at the Indian Statistical Institute, Bangalore and visiting North Dakota State University, Fargo.

The author acknowledges the efforts of Mr. Dharmappa of ISIBC for word-processing this monograph.

Dr. Surekha Rao, my wife, in spite of her own busy academic work, has been a constant source of encouragement to me throughout the preparation of this monograph. I am full of appreciation for her.

K.P.S. Bhaskara Rao

# Chapter 1

# Elementary results on rings

We shall start with an introduction of various algebraic concepts required for this monograph. Most of the concepts are elementary and any good basic algebra text book is sufficient to get an understanding of these concepts. We need no deep results. We shall try to be as complete as possible with the details. For this reason this monograph can be understood by any one with a rudimentary knowledge of algebra and a rudimentary knowledge of matrices.

We shall quickly go through various definitions and give lot of examples.

A *Ring* (or an *Associative Ring*) $R$ with 1 is a non-vacuous set $R$ together with two binary operations $+$ and $\cdot$ on $R$ and two distinguished elements $0, 1$ in $R$ such that $(R, +, 0)$ is a commutative group, $(R, \cdot, 1)$ is a semigroup, 1 satisfies the condition that $1 \cdot x = x \cdot 1 = 1$ for all $x$ in $R$ and the distributive laws viz., $a(b + c) = ab + ac$ and $(b + c)a = ba + ca$ for all $a, b, c$ in $R$, hold. 1 shall be called the *identity* of $R$. When we talk about rings with 1 or rings with identity we mean the same. If $S$ is a nonempty subset of $R$, an element $a$ of $R$ shall be called an identity element for $S$ if $ax = xa = x$ for every $x$ in $S$. A *Ring with an involution* $a \to \bar{a}$ is a ring $R$ with a function $a \to \bar{a}$ from $R$ *to* $R$ such that $\overline{a+b} = \bar{a} + \bar{b}$, $\overline{ab} = \bar{b}\bar{a}, \bar{0} = 0$ and $\bar{1} = 1$.

A ring $R$ with 1 is called a *commutative ring* if $\cdot$ is commutative. Most of the time we simply write $ab$ for $a \cdot b$ ignoring the $\cdot$. A *subring* of a ring $R$ is a subset $S$ of $R$ so that $(S, +, \cdot)$ is a ring by itself with the 0 and 1 of $R$ being the 0 and 1 of $S$ also. An element $a$ of $R$ is called a *unit* if there is a $b$ in $R$ such that $ab = ba = 1$. If $a$ and $b$ are nonzero elements of a ring $R$ with 1 such that $a \cdot b = 0$ we call $a$ and $b$ *zero divisors.* An element $e$ of a ring $R$ is called an *idempotent* if $e^2 = e$. We say that $a$ *divides* $b$ and write $a|b$ if there is a $c$ in $R$ such that $a \cdot c = b$. Since we shall use the concept of $a$ divides $b$ in the case of commutative rings only there will be no confusion. If $a_1, a_2, \cdots, a_n \in R$ we say that $a$ is the *greatest common divisor* of $a_1, a_2, \cdots, a_n$ if $a|a_1, a|a_2, \cdots, a|a_n$ and $b|a_1, b|a_2, \cdots, b|a_n \Rightarrow b|a$ and we write $g.c.d.(a_1, a_2, \cdots, a_n) = a$.

We shall use some standard notations:

$\mathbb{Z}$ = all integers.

$F$ = a field.

$\mathbb{R}$ = the field of real numbers.

$\mathbb{C}$ = the field of complex numbers.

$R$ = a ring.

An *ideal* in a commutative ring $R$ is a nonempty subset $\mathcal{I}$ of $R$ such that $\mathcal{I}$ is closed under $+$ and for all $a \in \mathcal{I}$ and $b \in R$, $ab$ is also in $\mathcal{I}$. Of course, the singleton set $\{0\}$ as well as $R$ are ideals in $R$. If $\mathcal{I} \neq \{0\}$ we call $\mathcal{I}$ a nonzero ideal. Given $a_1, a_2, \cdots, a_n$ in $R$ we can generate an ideal by simply taking $\{b_1a_1 + b_2a_2 + \cdots + b_na_n : b_1, b_2, \cdots, b_n \in R\}$, to be denoted by $<a_1, \cdots, a_n>$. This is the smallest ideal containing the set $\{a_1, a_2, \cdots, a_n\}$ and is called the ideal generated by $a_1, a_2, \cdots, a_n$.

A commutative ring $R$ is called an *integral domain* if it has no zero divisors. If in an integral domain $R$ every ideal $\mathcal{I}$ is of the form $<a>$ for some $a \in R$ we shall call $R$, a *principal ideal ring* or a *principal ideal domain.*

$R$ is called a *Euclidean domain* (there are several definitions in the literature all of which essentially are equivalent) if there is a function $|\cdot| : R \to \mathbb{Z}^+$ such that

(i) $|a| = 0$ if and only if $a = 0$;

(ii) $|ab| \geq |a|$ for all nonzero $a$ and $b$; and

(iii) for $a \in R$ and $b \neq 0$ in $R$ there exist $c$ and $d$ in $R$ such that $a = cb + d$ and $|d| < |b|$.

$|\cdot|$ is called a *Euclidean norm.*

Now we shall give the examples and some elementary results.

1. The set of all integers to be denoted by $\mathbb{Z}$ with the usual $+, \cdot, 0$ and 1 is a commutative ring. It is also an integral domain, a Euclidean domain and so also is a principal ideal domain. It is not a field.

2. Every Euclidean domain is a principal ideal domain. In fact, if $(R, |\cdot|)$ is a Euclidean domain and if $\mathcal{I}$ is a nonzero ideal, then an element $a \in \mathcal{I}$ such that $|a| = \min\{|b| : b \in \mathcal{I}, b \neq 0\}$ will satisfy $\mathcal{I} = <a>$.

3. Not every principal ideal domain is a Euclidean domain.

4. If $a_1, a_2, \cdots, a_n \in R$, a principal ideal ring, then there is an $a \in R$ such that $<a_1, \cdots, a_n> = <a>$. This implies that the $g.c.d.(a_1, \cdots, a_n) = a$. Thus in a principal ideal ring every finite set of elements $a_1, a_2, \cdots, a_n$ have a $g.c.d.$ and this $g.c.d.$ can be written as a linear combination of $a_1, a_2, \cdots, a_n$. In $\mathbb{Z}$ we shall write $(m, n)$ for the $g.c.d.(m, n)$.

5. Every field is a Euclidean domain with a trivial Euclidean norm (trivially!). If $F$ is a field $F[x]$ stands for all the polynomials in the indeterminate $x$ with coefficients from $F$. $F[x]$ is also a Euclidean domain with the norm $|a_0 + a_1x + \cdots + a_nx^n| = n + 1$ for any polynomial $a_0 + a_1x + \cdots + a_nx^n$ with $a_n \neq 0$ and $|0| = 0$. $F[x]$ is not a field.

6. If $R$ is a commutative ring, $R[x_1, x_2, \cdots, x_n]$ stands for the commutative ring of all polynomials in the indeterminates $x_1, x_2, \cdots, x_n$ with coefficients from $R$. If $R$ is an integral domain, $R[x_1, \cdots, x_n]$ is also an integral domain. If $F$ is a field $F[x_1, x_2, \cdots, x_n]$ is an integral domain. If $n \geq 2, F[x_1, x_2, \cdots, x_n]$ is not a PID. That, for $n \geq 2, F[x_1, x_2, \cdots, x_n]$ is not a PID can be seen by considering the ideal generated by $x_1, \cdots, x_n$.

7. Every integral domain can be embedded in a field. This is a standard procedure akin to the relation of $\mathbb{Z}$ to $Q$. Every integral domain with an involution can be embedded in a field with an involution.

8. If $k \in \mathbb{Z}$ the set $\mathbb{Z}[k] = \{0, 1, \cdots, k-1\}$ with $+$ defined by $a + b = a + b(mod\ k)$ and $a \cdot b = ab(mod\ k)$ makes $\mathbb{Z}[k]$ a ring. $\mathbb{Z}[k]$ is a field if and only if $\mathbb{Z}[k]$ is an integral domain if and only if $k$ is a prime.

9. All real valued continuous functions defined on $[0, 1]$ is a commutative ring and it has zero divisors.

10. All rational functions $a(x_1, x_2, \cdots, x_n)b(x_1, x_2, \cdots, x_n)^{-1}$ where $a(x_1, x_2, \cdots, x_n)$ and $b(x_1, x_2, \cdots, x_n)$ are polynomials with real coefficients and such that $b(x_1, x_2, \cdots, x_n) \neq 0$ for all real $x_1, x_2, \cdots, x_n$ is a ring. This ring appears naturally in the theory of systems over rings. This ring will be denoted by $R[x_1, x_2, \cdots, x_n]^*$.

11. Quillen-Souslin Theorem says that every finitely generated projective module over the ring $F[x_1, x_2, \cdots, x_n]$, is free. In fact, for any principal ideal domain $R$, every finitely generated projective module over the ring $R[x_1, x_2, \cdots, x_n]$, is free. There are integral domains and finitely generated projective modules over them which are not free.

# Chapter 2

# Matrix algebra over rings

In this Chapter we shall develop the basic results in matrix algebra over rings.

## 2.1 Elementary notions

If $I$ and $J$ are index sets of finite cardinality, a map $\sigma : I \times J \to R$ will be called a matrix. If $i \in I$ and $j \in J$, then $\sigma(i,j)$ is the $(i,j)$th element of the matrix $\sigma$. If $I = \{1, 2, \cdots, m\}$ and $J = \{1, 2, \cdots, n\}$, we will simply write $A$ for a matrix with $(i,j)$th element being $a_{ij}$ and we write this as a rectangular $m \times n$ array and call $A$ an $m \times n$ matrix.

If $\sigma : I \times J \to R$ and $\tau : I \times J \to R$ are two matrices we write $\sigma + \tau$ for the sum of the two matrices defined by $(\sigma + \tau)(i,j) = \sigma(i,j) + \tau(i,j)$.

If $\sigma : I \times J \to R$ and $\tau : I \times J \to R$ are two matrices we write $\sigma \cdot \tau$ for the product of the two matrices defined by $(\sigma \cdot \tau)(i,k) = \sum\limits_{j \in J} \sigma(i,j)\tau(j,k)$. Thus $\sigma \cdot \tau : I \times K \to R$.

With the above definitions we can easily verify: $\sigma \cdot (\tau \cdot \eta) = (\sigma \cdot \tau) \cdot \eta$, $\sigma \cdot (\tau + \eta) = \sigma \cdot \tau + \sigma \cdot \eta$ and $(\tau + \eta) \cdot \sigma = \tau \cdot \sigma + \eta \cdot \sigma$ for matrices defined on appropriate index sets. We define the identity matrix $I : J \times J \to R$ to be the map defined by $I(j,j) = 1$ for all $j \in J$ and $I(j,k) = 0$ for $j \neq k$ in $J$. We also define the zero matrix $0 : J \times J \to R$ to be the map $0(j,k) = 0$ for all $j, k$ in $J$. By a diagonal matrix we mean a matrix $D = (d_{ij})$ such that $d_{ij} = 0$ for all $i \neq j$. For convenience we sometimes write a diagonal matrix $D$ as *diag* $(d_1, d_2, \cdots, d_r)$ to mean that $D$ is possibly an $m \times n$ matrix whose $(1,1), (2,2), \cdots, (r,r)$-elements are $d_1, d_2, \cdots, d_r$ respectively and all other elements are zero.

We shall freely use rectangular arrays for matrices, even though, some times our index sets are not of the form $\{1, 2, \cdots, n\}$. $I$ also stands for the square identity matrix.

For an $m \times n$ matrix $A$, the transpose of $A$ denoted by $A^T$ is an $n \times m$ matrix whose $(i, j)$th entry is the $(j, i)$th entry of $A$. Clearly $(A + B)^T = A^T + B^T, (AB)^T = B^T A^T$ and $(A^T)^T = A$. In case the ring $R$ (not necessarily commutative) has an involution $a \to \bar{a}$, for an $m \times n$ matrix $A = (a_{ij})$ we write $\overline{A}$ for the matrix whose $(i, j)$th element is $\bar{a}_{ij}$ and $A^*$ for $(\overline{A})^T$. Clearly $(A + B)^* = A^* + B^*, (AB)^* = B^* A^*$ and $(A^*)^* = A$.

For integers $m \geq 1$ and $1 \leq k \leq m$ we shall let $Q_{k,m}$ stand for the set $\{\alpha : \alpha = \{\alpha_1, \alpha_2, \cdots, \alpha_k\}, 1 \leq \alpha_1 < \alpha_2 < \cdots < \alpha_k \leq m\}$. For integers $m, n \geq 1$ and $1 \leq k \leq \min\{m, n\}$ we shall let $\Delta_{k,m,n}$ stand for the set $Q_{k,m} \times Q_{k,n}$. We shall write $\Delta$ for $\Delta_{k,m,n}$ if there is no scope for confusion.

If $A$ is an $m \times n$ matrix, $\alpha = \{\alpha_1, \alpha_2, \cdots, \alpha_k\}$ with $\alpha_1 < \alpha_2 < \cdots < \alpha_k$ is an element of $Q_{k,m}$ and $\beta = \{\beta_1, \beta_2, \cdots, \beta_\ell\}$ with $\beta_1 < \beta_2 < \cdots < \beta_\ell$ is an element of $Q_{\ell,n}$ then $A_{\alpha\beta}$ (sometimes written also as $A_{\alpha,\beta}$) stands for the $k \times \ell$ matrix whose $(i, j)$th element is the $(\alpha_i, \beta_j)$ element of $A$. $A_{\alpha\beta}$ is called the $(\alpha, \beta)$-submatrix of $A$. $B$ is called a submatrix of $A$ if there are $\alpha$ and $\beta$ such that $B = A_{\alpha\beta}$. If $\alpha = \{\alpha_1, \cdots, \alpha_k\}$, with $\alpha_1 < \alpha_2 < \cdots < \alpha_k$, is a subset of $\{1, 2, \cdots, m\}$ then $|\alpha|$ stands for the number of elements in $\alpha$ (in this case, $|\alpha| = k$) and $\overline{\alpha}$ stands for the complement of $\alpha$ $(= \{1, 2, \cdots, m\} - \{\alpha_1 \cdots \alpha_k\})$ written in increasing order. So $A_{\overline{\alpha}\overline{\beta}}$ stands for another submatrix of $A$. We shall need this notation while dealing with Laplace expansion theorem. If $\beta = \{1, 2, \cdots, n\}$, we shall write $A_{\alpha *}$ for $A_{\alpha\beta}$. Similarly for $\alpha = \{1, 2, \cdots, m\}$ we shall write $A_{*\beta}$ for $A_{\alpha\beta}$.

Observe that all $n \times n$ matrices over a ring $R$ with 1 forms a ring with 1, though not commutative in general. This ring will be denoted by $R_{n \times n}$. An $n \times n$ matrix $E$ is said to be idempotent if $E$ is an idempotent element of $R_{n \times n}$, i.e., $E^2 = E$. If $E$ is idempotent, trivially, $I - E$ is also idempotent.

The trace of an $n \times n$ matrix, to be denoted by $Tr(A)$, is defined as $\sum_{i=1}^{n} a_{ii}$. Clearly over a commutative ring, if $A$ is an $m \times n$ matrix and $B$ is an $n \times m$ matrix, $Tr(AB) = Tr(BA)$.

If $A$ is an $m \times n$ matrix, an $n \times m$ matrix $B$ is called a right inverse (left inverse) of $A$ if $AB = I$ $(BA = I)$. $B$ is called an inverse of $A$ if $B$ is both a right and a left inverse. If a matrix $A$ has an inverse then $A$ is a square matrix and the inverse must be unique. If $A$ is a square matrix over a commutative ring and if $A$ has a right inverse $B$ then $B$ must be the inverse of $A$. These results will follow from our results on determinants later.

We shall follow the usual notation about vectors as column vectors. We write $\underline{x} = (x_1, x_2, \cdots, x_n)^T$. For example, $\underline{x}^T A\underline{y}$ stands for $\sum\limits_{i,j} x_i a_{ij} y_j$; where $\underline{x} = (x_1, x_2, \cdots, x_n)^T$, $\underline{y} = (y_1, y_2, \cdots, y_n)^T$ and $A = (a_{ij})$.

Usual algebra of partitioned matrices also holds for matrices over rings (not necessarily commutative). We shall extensively use partitioned matrices. For example, if we write a partitioned matrix $\begin{bmatrix} A & B \\ C & D \end{bmatrix}$ it automatically implies that the matrices $A, B, C$ and $D$ are of appropriate order. Also, when we consider products of partitioned matrices we have to be extremely careful in making sure that the matrices are of appropriate order. Many times, if there is no confusion, we use partitioned matrices without explicitely verifying the orders.

Let us also describe some elementary operations on matrices and explain their relation to multiplication of matrices.

Firstly note that, if $A$ is an $m \times n$ matrix and $B$ is an $n \times k$ matrix, the rows of $AB$ can be described in terms of the rows of $B$ and columns of $AB$ can be described in terms of columns of $A$. If $(x_1, x_2, \cdots, x_n)$ is the first row of $A$, then, the linear combination of the rows of $B$ with $x_1, x_2, \cdots, x_n$ as coefficients (in that order) gives the first row of $AB$. A similar remark applies to the columns of $AB$.

Now, let $A = (a_{ij})$ be an $m \times n$ matrix over $R$. There are three types of elementary operations, which one can think of, on the rows of $A$: (i) multiplication of a particular row of $A$ by a nonzero element of $R$, (ii) addition of a multiple of a row to another row and (iii) interchange of two rows. As in the classical case, all these three operations can be achieved by premultiplying $A$ by a matrix $C$ of order $m \times m$ where $C$ is obtained from the identity matrix $I$ either (i) by multiplication of a particular row of $I$ by an element of $R$ or (ii) by addition of a multiple of a row of $I$ to another row of $I$ or (iii) by interchange of two rows of $I$. Column operations can also be described by postmultiplication in a similar way.

For example, let us consider the operation of the first row of an $m \times n$ matrix $A$ being replaced by a linear combination of the rows of $A$ with coefficients $x_1, x_2, \cdots, x_m$ (in that order). This operation can be achieved by considering the product matrix $CA$ where $C$ is an $m \times m$ matrix obtained from $I$ by replacing the first row of $I$ by $(x_1, x_2, \cdots, x_n)$.

## 2.2 Determinants

Let $A$ be an $n \times n$ matrix over a commutative ring $R$. The determinant of $A$, written as $|A|$, as in the classical case of real numbers, is defined by

$$|A| = \sum_{\sigma \in S_n} sgn(\sigma) a_{1\sigma(1)} a_{2\sigma(2)} \cdots a_{n\sigma(n)}$$

where $S_n$ stands for all the permutations of $\{1, 2, \cdots, n\}$ and $sgn(\sigma) = 1$ if $\sigma$ is an even permutation and $sgn(\sigma) = -1$ if $\sigma$ is an odd permutation. Many of the classical results on determinants hold good for determinants over commutative rings also, with exactly the same arguments.

For example, $|I| = 1$ and $|0| = 0$. One can also show that $|AB| = |A||B|$ for $n \times n$ matrices $A$ and $B$. $|A^T| = |A|$ also holds. If $B$ is obtained by multiplying a particular row (or column) of $A$ by an element $c$, then $|B| = c|A|$. If $B$ is obtained by interchanging two rows (or two columns) of $A$, then $|B| = -|A|$. If two rows of $A$ are identical, then $|A| = 0$. If $A, B$ and $C$ are three $n \times n$ matrices with identical rows except for the $i$th row for a fixed $i$ and if the $i$th row of $A$ is the sum of the $i$th rows of $B$ and $C$ then $|A| = |B| + |C|$. A similar statement for the columns also holds.

The determinant of a square submatrix of a matrix $A$ will be called a minor of $A$.

Two important Theorems from the classical matrix algebra also hold for matrices over commutative rings. These play an important role in this monograph. They are,

**Theorem 2.1: Laplace Expansion Theorem.**

Let $A$ be an $n \times n$ matrix over a commutative ring.

(a) For any fixed $i$, $|A| = \sum_j (-1)^{i+j} a_{ij} |A_{\overline{ij}}|$ (Expansion of $|A|$ by the $i$th row).

and also,

for any fixed $j$, $|A| = \sum_i (-1)^{i+j} a_{ij} |A_{\overline{ij}}|$ (Expansion of $|A|$ by the $j$th column).

(b) If $\alpha$ is a fixed subset of $\{1, 2, \cdots, n\}$ then

$$|A| = \sum_{\beta \subset \{1,2,\cdots,n\}, |\beta| = |\alpha|} (-1)^{S(\alpha)+S(\beta)} |A_{\alpha\beta}||A_{\overline{\alpha}\overline{\beta}}|$$

(Expansion of $|A|$ by the rows in $\alpha$)

and also,
for any fixed subset $\beta$ of $\{1, 2, \cdots, n\}$,

$$|A| = \sum_{\alpha \subset \{1,2,\cdots,n\}, |\alpha|=|\beta|} (-1)^{S(\alpha)+S(\beta)} |A_{\alpha\beta}||A_{\overline{\alpha}\overline{\beta}}|$$

(Expansion of $|A|$ by the columns in $\beta$.)

Here if $\alpha = \{\alpha_1, \alpha_2, \cdots, \alpha_k\}$, $S(\alpha)$ stands for $\sum_{i=1}^{k} \alpha_i$ and $|\alpha| = k$.

♠

The second Theorem is

**Theorem 2.2: Cauchy-Binet Theorem.**

Let $m \leq n$. If $A$ is an $m \times n$ matrix and $B$ is an $n \times m$ matrix over a commutative ring.

$$|AB| = \sum_{\beta:|\beta|=m,\ \beta \subset \{1,2,\cdots,n\}} |A_{*\beta}||B_{\beta*}|.$$

♠

The classical proofs of the above two Theorems go through in the case of commutative rings also.

Part (a) of the Laplace Expansion Theorem can be used to derive the following result. If $A$ is an $n \times n$ matrix, we define the adjoint of $A$, denoted by $adj(A)$ to be the $n \times n$ matrix whose $(i, j)$th element is $(-1)^{i+j}|A_{\overline{ji}}|$. The Laplace Expansion Theorem can be used to derive the result: $Aadj(A) = |A|I$. The Laplace Expansion Theorem can also be used to prove that for a partitioned matrix of the form $\begin{bmatrix} A & 0 \\ B & C \end{bmatrix}$, where $A$ and $C$ are square matrices, its determinant $\begin{vmatrix} A & 0 \\ B & C \end{vmatrix} = |A||C|$.

## 2.3 The $\frac{\partial}{\partial a_{ij}}$ notation

If $A = (a_{ij})$ is an $n \times n$ matrix, for a fixed $i$ and a fixed $j$ $\frac{\partial |A|}{\partial a_{ij}}$ stands for the coefficient of $a_{ij}$ in the expansion of the determinant of $A$, i.e., $\frac{\partial |A|}{\partial a_{ij}} = (-1)^{i+j}|A_{\overline{ij}}|$. Sometimes we shall also write $\frac{\partial}{\partial a_{ij}}|A|$ for $\frac{\partial |A|}{\partial a_{ij}}$.

Now, part (a) of the Laplace Expansion Theorem can be restated as: $|A| = \sum_j a_{ij} \frac{\partial}{\partial a_{ij}}|A|$. Also, $\sum_j c_j \frac{\partial}{\partial a_{ij}}|A|$ is equal to $|B|$ where $B$ is obtained from $A$ by replacing the $i$th row of $A$ by the row of $c_j$'s. Also, if $\underline{x} = (x_1, x_2, \cdots, x_n), \underline{y} = (y_1, y_2, \cdots, y_n)^T$, $A$ is an $n \times n$ matrix and $c$ is an element of the ring, then

$$\begin{vmatrix} A & \underline{x} \\ \underline{y}^T & c \end{vmatrix} = c|A| - \sum_{i,j} x_i y_j \frac{\partial}{\partial a_{ij}} |A| = c|A| - \underline{y}^T adj(A)\underline{x}.$$

If $A = (a_{ij})$ is an $m \times n$ matrix and $A_{\alpha\beta}$ is a submatrix of $A$ with $|\alpha| = |\beta|$ and $a_{ij}$ is an element of $A$ then we define $\frac{\partial}{\partial a_{ij}}|A_{\alpha\beta}|$ as above if $i \in \alpha$ and $j \in \beta$ and $= 0$ if $i \notin \alpha$ or $j \notin \beta$.

Usual rules of partial derivatives apply. For example: $\frac{\partial}{\partial a_{ij}} \sum |A_{\alpha\beta}| c_{\alpha\beta}$ stands for $\sum \frac{\partial |A_{\alpha\beta}|}{\partial a_{ij}} c_{\alpha\beta}$.

If $A = BC$ where $B$ is an $m \times k$ matrix and $C$ is a $k \times n$ matrix (with $k \leq n$), $\alpha$ and $\beta$ are sets of indices with $|\alpha| = |\beta| = k$ and $i \leq m$ and $j \leq n$ are any two indices then $\frac{\partial}{\partial a_{ij}}|A_{\alpha\beta}| = \sum_{\ell=1}^{k} \frac{\partial}{\partial b_{i\ell}}|B_{\alpha *}| \frac{\partial}{\partial c_{\ell j}}|C_{*\beta}|$. This follows from the Cauchy-Binet Theorem.

Similar to the definition of $\frac{\partial}{\partial a_{ij}}|A_{\alpha\beta}|$ we also define, for $(\gamma, \delta) \in \Delta_{k,m,n}$ $\frac{\partial}{\partial |A_{\gamma\delta}|}|A_{\alpha\beta}|$ to be $(-1)^{S(\alpha)+S(\beta)}|A_{\overline{\alpha}\overline{\beta}}|$ where $S(\alpha) + S(\beta)$ is defined as in Laplace Expansion Theorem.

We shall extensively use the $\frac{\partial}{\partial a_{ij}}$ notation in this monograph. One should note that even if $|A| = 0$, $\frac{\partial}{\partial a_{ij}}|A|$ need not be equal to 0.

## 2.4 Rank of a matrix

For an $m \times n$ nonzero matrix $A$ over a commutative ring the largest $k \geq 0$ for which there exist $\alpha$ and $\beta$ with $|\alpha| = |\beta| = k$ such that $|A_{\alpha\beta}| \neq 0$ will be called the rank of the matrix $A$. We define the rank of the zero matrix to be zero. We shall denote the rank of the matrix $A$ by $\rho(A)$.

This rank, which is also called the determinantal rank, has most of the properties of the usual rank. Let us summarize them. The proof of the following Theorem is clear. One has to use Theorem 2.2 for proving (d).

**Theorem 2.3:** Let $A \in R_{m \times n}$. Then

(a) $0 \leq \rho(A) \leq \min\{m, n\}$.

(b) $\rho(A) = \rho(A^T)$.

(c) $\rho(A) = 0$ if and only if $A = 0$.

(d) $\rho(AB) \leq \min\{\rho(A), \rho(B)\}$, if $B \in R_{n\times k}$.
(e) $\rho(CAB) = \rho(A)$ if $C$ has a left inverse and $B$ has a right inverse.

♠

This rank in general does not satisfy the property that $\rho\left(\begin{bmatrix} A & 0 \\ 0 & C \end{bmatrix}\right) = \rho(A) + \rho(C)$. This property holds for integral domains only. This rank also differs from the rank (called the McCoy rank below) defined in Chapter 4 of [58]. Our rank also has no relation to linear independence. For example, $\begin{bmatrix} 3 & 0 \\ 0 & 4 \end{bmatrix}$ has rank 1 over the ring $\mathbb{Z}[12]$.

In the literature there are other definitions of rank. Let us look at some of them briefly. Over a commutative ring with 1 one can define two other (rather three) concepts of ranks:

$c(A)$, called the column rank of $A$, is defined by $c(A)$ = the least cardinality of a generating set for the module generated by the columns of $A$.

$r(A)$, the row rank of $A$ is also defined in a similar way.

The Schein rank or the semiring rank of an $m \times n$ nonzero matrix $A$ is defined as the least integer $k$ for which there are matrices $B$ and $C$ of order $m \times k$ and $k \times n$ respectively such that $A = BC$. This definition is derived from the fact that every matrix over a field can be factorized as $BC$ where $B$ is of order $m \times k$, $C$ is of order $k \times n$ and $k = \rho(A)$ and $\rho(A)$ is the smallest such $k$. We shall denote this rank by $S(A)$ – for Boris M. Schein.

The McCoy rank of an $m \times n$ nonzero matrix $A$ is defined as the largest integer $t$ such that $Annihilator(\mathcal{I}_t(A)) = 0$ where $Annihilator(\mathcal{I}_t(A)) = \{x \in R : x|B| = 0 \text{ for every } t \times t \text{ submatrix } B \text{ of } A\}$.

For the zero matrix all the ranks are defined to be zero.

One should note that the Schein rank does not demand that in $A = BC$ $B$ and $C$ need have left and right inverses respectively.

Most of the problems for defining a rank arise when the ring has zero divisors. The best possible rank which takes care of such rings is the McCoy rank. See McCoy[57] and McDonald[58] for more on this rank.

## 2.5 Compound matrices

Let $A$ be an $m\times n$ matrix and let $k \leq \min\{m, n\}$. The $k$th compound matrix of $A$, to be denoted by $C_k(A)$ is the $\binom{m}{k} \times \binom{n}{k}$ matrix whose $(\alpha, \beta)$th entry for any $\alpha = \{\alpha_1, \alpha_2, \cdots, \alpha_k\} \subset \{1, 2, \cdots, m\}$ with $\alpha_1 < \alpha_2 < \cdots < \alpha_k$ and $\beta = \{\beta_1, \beta_2, \cdots, \beta_k\} \subset \{1, 2, \cdots, n\}$ with $\beta_1 < \beta_2 < \cdots < \beta_k$ is $|A_{\alpha\beta}|$. The compound matrix $C_k(A)$ is simply the matrix of $k \times k$ minors of $A$.

Clearly, if $\rho(A) = r$ then $C_r(A) \neq 0$. Also, if A is an $n \times n$ matrix $C_{n-1}(A) = adj(A)$.

Compound matrices have many nice properties. Let us start with,

**Theorem 2.4:** For matrices over a commutative ring $R$,

(a) If $A$ is any matrix, $\rho(A) = r$ if and only if $C_s(A) \neq 0$ for all $s \leq r$ and $C_s(A) = 0$ for all $s > r$ (which is also equivalent to saying that $C_r(A) \neq 0$ and $C_s(A) = 0$ for all $s > r$).

(b) If $A$ and $B$ are matrices for which $AB$ is defined, then for every $k$,

$$C_k(AB) = C_k(A)C_k(B).$$

**Proof:** From the Laplace Expansion Theorem it follows that whenever $B$ is an $m \times m$ matrix such that for some $r \leq m$ all the $r \times r$ subdeterminants are zero then, $|B| = 0$. (a) follows from this result, (b) can be seen from the Cauchy-Binet Theorem by observing that $(AB)_{\alpha\beta} = A_{\alpha *}B_{*\beta}$.

♠

In case $A$ is a matrix over a commutative ring $R$ with an involution $a \to \bar{a}$ then $C_k(A^*) = (C_k(A))^*$ and $C_k(\overline{A}) = \overline{C_k(A)}$. The following result is crucial to this monograph. It calculates the rank of $C_r(A)$ when $\rho(A) = r$.

**Theorem 2.5:** Over a commutative ring, if $A$ is a matrix such that $\rho(A) = r$ then $\rho(C_r(A)) = 1$.

**Proof:** Since $\rho(A) = r$, every $(r+1) \times (r+1)$ minor of $A$ is zero and there is an $r \times r$ minor of $A$ which is nonzero. Hence $C_r(A)$ is a nonzero matrix. If we show that every $2 \times 2$ minor of $C_r(A)$ is zero, it will follow that $\rho(C_r(A)) = 1$.

Let $\alpha, \gamma$ be sets of row indices and $\beta, \delta$ be sets of column indices such that $|\alpha| = |\beta| = |\gamma| = |\delta| = r$. Consider the partitioned matrix.

$$T = \begin{bmatrix} A_{\alpha\beta} & A_{\alpha\delta} \\ A_{\gamma\beta} & A_{\gamma\delta} \end{bmatrix}.$$

This is a $2r \times 2r$ matrix. Though this is not a submatrix of $A$, its determinantal rank is equal, trivially, to the determinantal rank of the submatrix $A_{\alpha\cup\gamma,\beta\cup\delta}$ of $A$ where $\alpha \cup \gamma$ is the set theoretical union of $\alpha$ and $\gamma$ and $\beta \cup \delta$ is the set theoretical union of $\beta$ and $\delta$. Hence every $(r+1) \times (r+1)$ minor of $T$ is zero.

Let $A_{\alpha\beta} = B, A_{\alpha\delta} = C, A_{\gamma\beta} = D$ and $A_{\gamma\delta} = E$ so that

$$T = \begin{bmatrix} B & C \\ D & E \end{bmatrix}.$$

Let $\underline{b}_j$ be the $j$th row of $B$ and $\underline{c}_j$ be the $j$th row of $C$.

Observe that the determinant

$$\begin{vmatrix} \underline{b_1} & \underline{c_1} \\ \underline{b_2} & \underline{0} \\ \cdots & \cdots \\ \underline{b_r} & \underline{0} \\ D & E \end{vmatrix} = \begin{vmatrix} \underline{b_1} & \underline{0} \\ \underline{b_2} & \underline{0} \\ \cdots & \cdots \\ \underline{b_r} & \underline{0} \\ D & E \end{vmatrix} + \begin{vmatrix} \underline{0} & \underline{c_1} \\ \underline{b_2} & \underline{0} \\ \cdots & \cdots \\ \underline{b_r} & \underline{0} \\ D & E \end{vmatrix}$$

By expanding the left side determinant using Laplace Expansion with respect to the $r+1$ rows, namely, the first row and the last $r$ rows and using the fact that all $(r+1)\times(r+1)$ minors of $T$ are zero, we get that the left side determinant is equal to zero. Thus

$$\begin{vmatrix} B & 0 \\ D & E \end{vmatrix} = -\begin{vmatrix} \underline{0} & \underline{c_1} \\ \underline{b_2} & \underline{0} \\ \cdots & \cdots \\ \underline{b_r} & \underline{0} \\ D & E \end{vmatrix}.$$

Similarly,

$$\begin{vmatrix} \underline{0} & \underline{c_1} \\ \underline{b_2} & \underline{0} \\ \underline{b_3} & \underline{0} \\ \cdots & \cdots \\ \underline{b_r} & \underline{0} \\ D & E \end{vmatrix} = -\begin{vmatrix} \underline{0} & \underline{c_1} \\ \underline{0} & \underline{c_2} \\ \underline{b_3} & \underline{0} \\ \cdots & \cdots \\ \underline{b_r} & \underline{0} \\ D & E \end{vmatrix}.$$

Continuing in this way and combining all the equalities we get that

$$\begin{vmatrix} B & 0 \\ D & E \end{vmatrix} = (-1)^r \begin{vmatrix} 0 & C \\ D & E \end{vmatrix}$$

which in turn equals

$$(-1)^r(-1)^r \begin{vmatrix} C & 0 \\ E & D \end{vmatrix}.$$

Thus $|B|\cdot|E| = |C|\cdot|D|$. Thus every $2\times 2$ minor of $C_r(A)$ is zero. Thus $\rho(C_r(A)) = 1$.

♠

Later we shall need a finer version of part of the proof of the above Theorem. We shall isolate this and present it below.

**Proposition 2.6:** Let

$$T = \begin{bmatrix} B & C \\ D & E \end{bmatrix}$$

where $B$ is an $s \times s$ matrix, $E$ is a $t \times t$ matrix, $\rho(T) = r$ and $r \leq t$. Then

$$\begin{vmatrix} B & 0 \\ D & E \end{vmatrix} = (-1)^s \begin{vmatrix} 0 & C \\ D & E \end{vmatrix}.$$

**Proof:** The proof is no different from the proof of Theorem 2.5.

♠

## EXERCISES

**Exercise 2.1:** Compare the ranks $\rho(A)$, $c(A)$, $r(A)$ and $S(A)$ for matrices over various classes of rings. For example, they are all equal for matrices over fields. Study these ranks for matrices, for example, over Euclidean domains, principal ideal domains, integral domains, Boolean algebras and other interesting rings. See [11] and [16] for some information.

**Exercise 2.2:** Show that $\rho(C_k(A)) = \binom{r}{k}$ for $1 \leq k \leq r$ where $r$ is the rank of $A$.

**Exercise 2.3:** Is the converse of Theorem 2.5 true in general commutative rings? Over fields show that the converse of Theorem 2.5 is true.

**Exercise 2.4:** Show the result of Proposition 2.6 for matrices over an integral domain in a simpler way by using the fact that every integral domain can be embedded in a field (see [20]). See also [62] for another proof.

# Chapter 3

# Regular elements in a ring

In this Chapter we shall introduce regular elements in a ring and also study regular matrices. Though the concept of regular element in a ring is much older than the Moore-Penrose equations we shall first introduce the Moore-Penrose equations.

## 3.1 The Moore-Penrose equations

Let $R$ be a ring (not necessarily commutative) with 1 and an involution $a \to \bar{a}$. Recall that for any matrix $A$, $A^*$ stands for $(\overline{A})^T$.

Now, for matrices $A$ (of order $m \times n$) and $G$ (of order $n \times m$) over R we consider the Moore-Penrose equations:

$$AGA = A \tag{1}$$
$$GAG = G \tag{2}$$
$$(AG)^* = AG \tag{3}$$
$$(GA)^* = GA \tag{4}$$

If $A$ and $G$ satisfy (1) then $G$ will be called a generalized inverse of $A$ (or a $g$-inverse of $A$ or a 1-inverse of $A$ or a regular inverse of $A$ or an inner inverse of $A$). We denote an arbitrary $g$-inverse of $A$ by $A^-$ if there is one. A matrix $A$ is said to be regular if it has a $g$-inverse.

If $A$ and $G$ satisfy (2) then $G$ will be called a 2-inverse of $A$ or an *outer inverse* of $A$.

If $A$ and $G$ satisfy both (1) and (2) then $G$ will be called a reflexive $g$-inverse of $A$ (or a $\{1,2\}$-inverse of $A$). Every regular matrix has a reflexive $g$-inverse. In fact, if $G_1$ and $G_2$ are $g$-inverses of $A$ then $G_1AG_2$ is a reflexive

$g$-inverse of $A$. Also, not every $g$-inverse is a reflexive $g$-inverse (even in the case of real matrices).

If $A$ and $G$ satisfy all the equations (1) to (4) then $G$ is called a Moore-Penrose inverse of $A$. We shall write M-P inverse for Moore-Penrose inverse. We denote a Moore-Penrose inverse of $A$ by $A^+$. Over an integral domain if a matrix $A$ has a Moore-Penrose inverse, it is unique. Over a commutative ring also the Moore-Penrose inverse when it exists is unique. This can be seen as follows:

Let $G_1$ and $G_2$ be both Moore-Penrose inverses of a matrix $A$. Then, $G_1 = G_1AG_1 = A^*G_1^*G_1 = A^*G_2^*A^*G_1^*G_1 = A^*G_2^*G_1AG_1 = A^*G_2^*G_1 = G_2AG_1 = G_2AG_2AG_1 = G_2(G_2^*A^*)(G_1^*A^*) = G_2G_2^*A^* = G_2AG_2 = G_2$.

If $A$ and $G$ satisfy (1) and (3) then $G$ is called $\{1,3\}$-inverse of $A$ and is denoted by $A_{13}^-$. Similarly, if $A$ and $G$ satisfy (1) and (4) then $G$ is called $\{1,4\}$-inverse of $A$ and is denoted by $A_{14}^-$. If $A$ has $\{1,3\}$-inverse then it has a $\{1,2,3\}$-inverse. This can be seen by verifying that if $G$ is $\{1,3\}$-inverse of $A$ then $GAG$ is $\{1,2,3\}$-inverse of $A$. Also note that if $G$ is $\{1,3\}$-inverse of $A$ and $H$ is $\{1,4\}$-inverse of $A$ then $HAG$ is the Moore-Penrose inverse of $A$. $\{1,3\}$-inverses and $\{1,4\}$-inverses have great significance in the case of real matrices and also in the case of matrices over formally real fields (see the Exercises at the end of Chapter 4).

We also consider the following equations applicable to square matrices

$$AG = GA \tag{5}$$

$$A^k = A^{k+1}G \tag{$1^k$}$$

for an integer $k \geq 1$.

If $A$ and $G$ satisfy (1), (2) and (5) then $G$ is called a *group inverse of* $A$. We denote a group inverse of $A$ by $A^{\#}$. If $A$ and $G$ satisfy (1) and (5) only, then $G$ is called a *commuting* $g$-inverse of $A$.

If $A$ and $G$ satisfy (2), (5) and ($1^k$) for some $k \geq 1$ then $G$ is called a *Drazin inverse of* $A$.

There are other types of $g$-inverses which we shall be looking at. We shall define them as and when it becomes necessary.

## 3.2 Regular elements and regular matrices

Let $R$ be a ring, not necessarily commutative. Recall that an element $a \in R$ is said to be a unit if it has an inverse, i.e., if there is an element $a^{-1} \in R$ such that $a \cdot a^{-1} = a^{-1} \cdot a = 1$. If $a$ is a unit of $R$ and if $d$ is another element of $R$, then a solution to $ax = d$ exists and $x = a^{-1}d$ is a solution. Similarly, $ya = d$ is also solvable and $y = da^{-1}$ is a solution.

Suppose now that $a \in R$ is not a unit. If we demand that $ax = d$ be solvable for every $d \in R$, then demanding this for $d = 1$ gives rise to the concept of a right inverse of $a$. Let us say that an $a_R^{-1} \in R$ is a right inverse of $a$ if $aa_R^{-1} = 1$. Now, if $a \in R$ has a right inverse, then, for every $d \in R, ax = d$ has a solution and $x = a_R^{-1}d$ is a solution. Also, if $a \in R$ has a right inverse, then, $ya = d$ may not have a solution for every $d \in R$, but, if $ya = d$ has a solution, then, $a_R^{-1}$ can be used to find a solution. This is because $da_R^{-1}a = yaa_R^{-1}a = ya = d$, showing that $y = da_R^{-1}$ is a solution of $ya = d$ if it is solvable. Similarly the left inverse also plays its role.

Now, if we do not demand that $ax = d$ is solvable for every $d \in R$, but that we are looking for the existence of an element $g \in R$ such that whenever $ax = d$ is solvable, $x = gd$ is a solution, then we have the following Theorem.

**Theorem 3.1:** Let $R$ be an associative ring with 1. Let $a \in R$ and $g \in R$. Then the following are equivalent:

(i) Whenever $d \in R$ is such that $ax = d$ has a solution, $x = gd$ is a solution of $ax = d$.

(ii) $aga = a$.

(iii) Whenever $d \in R$ is such that $ya = d$ has a solution, $y = dg$ is a solution of $ya = d$.

**Proof:** (i) $\Rightarrow$ (ii). Clearly, since $R$ has 1, $ax = a$ has a solution. Hence $aga = a$.

(ii) $\Rightarrow$ (i). If $d \in R$ is such that $ax = d$ has a solution then $agax = ax = d$, from (ii), i.e., $agd = d$ or $x = gd$ is a solution.

(ii) $\Leftrightarrow$ (iii) can be shown in an analogous way.

♠

This gives rise to the definition: Let us say that an element $a$ in a ring $R$ with 1 is *regular* if there is an element $g \in R$ such that $aga = a$. We shall call $g$ a *regular inverse* or *generalized inverse* of $a$. We shall use the notation $a^-$ for a regular inverse of $a$ if it exists.

Clearly every unit in $R$ is regular. Also every $a \in R$ which has a left inverse or a right inverse is also regular. In the ring of integers $0, 1, -1$ are the only regular elements. In the ring of real numbers every element is regular. More generally, in any field every element is regular. The central question we would be interested in is to characterize regular elements in specific rings.

If $R$ is an associative ring without 1 the above Theorem is no longer valid. The ring of all $2 \times 2$ real matrices of the type $\begin{bmatrix} 0 & a \\ 0 & 0 \end{bmatrix}$ is a ring in which (i) of the above Theorem holds but (ii) does not hold. We shall state the following Theorem for academic interest. We shall not pursue the idea of the Theorem further. Subsequently we shall deal with rings with identity only.

**Theorem 3.2:** Let $R$ be an associative ring with or without 1, not necessarily commutative. Let $a \in R$ and $g \in R$. Then the following are equivalent.

(i) Whenever $d \in R$ is such that $ax = d$ has a solution, $x = gd$ is a solution of $ax = d$.

(ii) $agax = ax$ for all $x \in R$.

If $R$ has 1 or $R$ has no zero divisors then (ii) is equivalent to

(iii) $aga = a$.

♠

Hence, in an associative ring $R$ with or without identity we shall say that $a$ is *right regular* (*left regular*) if there is an element $g \in R$ such that $agax = ax$ for all $x \in R$ ($xaga = xa$ for all $x \in R$). For further analysis of these ideas see the Exercises at the end of this Chapter.

If $a$ is regular and $g$ is a regular inverse of $a$ then $ag$ and $ga$ are clearly idempotent. The following Theorem is clear.

**Theorem 3.3:** Let $R$ be a ring with 1. Let $a, g \in R$. Then $aga = a$ if and only if $ag$ is idempotent and there is an $x$ in $R$ such that $agx = a$. Also, $aga = a$ if and only if $ga$ is idempotent and there is a $y \in R$ such that $yga = a$.

♠

Now let us define regular matrices. An $n \times n$ matrix $A$ over a ring $R$ with 1 (not necessarily commutative) is called a regular matrix if $A$ is a regular element of the ring $R_{n \times n}$. If $A$ is an $m \times n$ matrix, we say that $A$ is a regular matrix if there is an $n \times m$ matrix $G$ over $R$ such that

$$AGA = A \tag{1}$$

If $A$ is regular we shall also say that $A$ has a generalized inverse and we shall call $G$ of (1) a *generalized inverse* of $A$. We shall abbreviate generalized inverse as $g$-inverse. The development in this section highlights the importance of generalized inverses of matrices.

Certain rank conditions are satisfied by generalized inverses of matrices. These are listed below.

**Theorem 3.4:** Let $A$ be an $m \times n$ matrix and $G$ be an $n \times m$ matrix over a commutative ring $R$. Then

(a) If $AGA = A$ then $\rho(A) \le \rho(G)$.

(b) If $GAG = G$ then $\rho(G) \le \rho(A)$.

(c) If $AGA = A$ and $GAG = G$ then $\rho(A) = \rho(G)$.

(d) If $AGA = A$ and $\rho(A) = \rho(G)$ then $GAG = G$ holds good if $R$ is an integral domain and does not hold good over general commutative rings.

**Proof:** (a), (b) and (c) are clear. The first part of (d) follows because such a result is true over any field. For the second part of (d), note that over the commutative ring $\mathbb{Z}[12]$ the ring of integers modulo 12, $4.1.4 = 4$ but $1.4.1 \neq 1$.

♠

It is well known that every real matrix has a $g$-inverse. More generally, every matrix over a field has a $g$-inverse. A more general statement is the Theorem of Von Neumann which we shall prove in the next section.

## 3.3 A Theorem of Von Neumann

As we have noted earlier, in some rings every element is a regular element. This is clearly so in the case of fields or division rings. Let us say that a ring $R$ is a *regular ring* (or, *regular*) if every element of $R$ is regular.

It is well known that every $n \times n$ real or complex matrix is regular. More generally, every $n \times n$ matrix over any field is regular. (see [34], [63] and [66]). These results are in fact special cases of a 1936 result of Von Neumann.

**Theorem 3.5 (Von Neumann):** If $R$ is a regular ring (not necessarily commutative and with or without 1) and $n \geq 1$ then the ring $R_{n\times n}$ of all $n \times n$ matrices over $R$ is also a regular ring.

Also, every $m \times n$ matrix over a regular ring is regular.

**Proof (Brown and McCoy):** We shall give the proof in several steps.

1. Observe that if $a$ and $y$ are such that $a - aya$ is regular then $a$ is regular. In fact, if $(a-aya)g(a-aya) = a-aya$, by expanding the left side and taking $aya$ from the right to the left we get that $a[g - yag - gay + yagay + y]a = a$. Hence $a$ is regular.

2. If every $2 \times 2$ matrix of the form $\begin{bmatrix} g & 0 \\ h & i \end{bmatrix}$ is regular then every $2 \times 2$ matrix is regular. In fact, if $\begin{bmatrix} a & b \\ c & d \end{bmatrix} = A$ is a $2 \times 2$ matrix by letting $Y = \begin{bmatrix} 0 & 0 \\ b^- & 0 \end{bmatrix}$ where $b^-$ is a regular inverse of $b$, we observe that $A - AYA$ is of the form $\begin{bmatrix} g & 0 \\ h & i \end{bmatrix}$ for some $g, h$ and $i$ in $R$. An application of (1) above gives the result.

3. If every $2 \times 2$ matrix of the from $\begin{bmatrix} 0 & 0 \\ f & 0 \end{bmatrix}$ is regular then every $2 \times 2$ matrix of the form $\begin{bmatrix} g & 0 \\ h & i \end{bmatrix}$ is regular. In fact, if $A = \begin{bmatrix} g & 0 \\ h & i \end{bmatrix}$ is a $2 \times 2$ matrix, by letting $Y = \begin{bmatrix} g^- & 0 \\ h & i^- \end{bmatrix}$ where $g^-$ is a regular inverse of $g$ and $i^-$ is a regular inverse of $i$, which exist because $R$ is regular, we observe that $A - AYA$ is of the form $\begin{bmatrix} 0 & 0 \\ f & 0 \end{bmatrix}$ for some $f$. An application of (1) above gives the result.

4. Now we shall show that if $R$ is a regular ring then every $2 \times 2$ matrix over $R$ is regular, i.e., $R_{2\times2}$ is a regular ring. In fact, for any $2 \times 2$ matrix of the form $A = \begin{bmatrix} 0 & 0 \\ f & 0 \end{bmatrix}$ the matrix $G = \begin{bmatrix} 0 & f^- \\ 0 & 0 \end{bmatrix}$ is a regular inverse of $A$. Combining (1), (2) and (3) we see that every $2 \times 2$ matrix over $R$ is regular.

5. $R_{4\times4}$ is a regular ring. In fact, $R_{4\times4}$ can be considered as $(R_{2\times2})_{2\times2}$. (4) above gives the result.

6. $R_{2^k \times 2^k}$ is a regular ring for all $k \geq 0$. This is similar to (5).

7. If $n \geq 1$, $R_{n\times n}$ is a regular ring. In fact, if $A$ is an $n \times n$ matrix over $R$ let $k$ be an integer such that $n \leq 2^k$. Let $B = \begin{bmatrix} A & 0 \\ 0 & 0 \end{bmatrix}$ where the three 0 matrices are of suitable orders so that the matrix $B$ is of order $2^k \times 2^k$. Let $H$ be a regular inverse of $B$. Partition $H$ as $\begin{bmatrix} H_1 & H_2 \\ H_3 & H_4 \end{bmatrix}$ according to the partition of $B$. Then one sees that $BHB = B$ gives that $AH_1A = A$ i.e., $A$ is regular.

8. If $A$ is an $m \times n$ matrix and $m \neq n$ exactly the same technique as in (7) shows that $A$ is regular. The $H_1$ of (7). will be an $n \times m$ matrix which is a $g$-inverse of $A$.

♠

## 3.4 Inverses of matrices

In a ring $R$ with 1, every unit of $R$ is clearly regular. In fact, if $a \in R$ is a unit, $a^{-1}$ is a regular inverse of $a$. If an $a \in R$ has a right inverse $a_R^{-1}$ (or left inverse $a_L^{-1}$ ) then also $a$ is regular. In fact, $a_R^{-1}(a_L^{-1})$ is a regular inverse of $a$, because $aa_R^{-1}a = a$ $(aa_L^{-1}a = a)$.

An $n \times n$ matrix $A$ over a commutative ring $R$ with 1 is called a *unimodular matrix* or an *invertible matrix* if it is a unit in the ring $R_{n\times n}$. The Identity matrix $I$ is trivially a unimodular matrix. There are other standard unimodular matrices – the matrices corresponding to the elementary operations of type (ii) and (iii). These are: the matrix obtained by adding a multiple of a particular row (or column) of $I$ to another row of $I$ and the matrix obtained by interchange of two rows of $I$. Product of unimodular matrices is again a unimodular matrix because $(AB)^{-1} = B^{-1}A^{-1}$. To find out about all unimodular matrices we have the following Theorem.

**Theorem 3.6:** An $n \times n$ matrix $A$ over a commutative ring $R$ with 1 is unimodular if and only if $|A|$ is a unit of $R$.

**Proof:** If $A$ is a unit of $R_{n\times n}$ then $AA^{-1} = I$ gives us that $|A||A^{-1}| = 1$, i.e., $|A|$ is a unit of $R$.

If $|A|$ is a unit of $R$ since $Aadj(A) = |A|I$, we get that $A[(|A|^{-1})adj(A)] = I$. Thus $A$ is invertible in $R_{n\times n}$.

♠

It follows from the discussion at the beginning of this section that if a matrix $A$ over a ring with 1 has an inverse then the inverse is a $g$-inverse. Over an integral domain $R$ a sort of converse is also true. If a matrix $A$ is such that $|A| \neq 0$ and $A$ has a $g$-inverse then the matrix $A$ is in fact unimodular and the $g$-inverse is the inverse. One can see this as follows: Let $AGA = A$. Hence $|A||G||A| = |A|$. Since $|A| \neq 0$ we get that $|G||A| = 1$. Thus $|A|$ is a unit in $R$. Thus $A$ is unimodular and hence $GA = I$. Thus $G$ is the inverse of $A$.

Recall from section 2.1 that an $m\times n$ matrix is said to have a right inverse (left inverse) if there is an $n \times m$ matrix $B$ such that $AB = I$ $(BA = I)$. When does an $m \times n$ matrix $A$ have a right inverse (or a left inverse)? A characterization in terms of determinants is given below. We shall also describe a method of finding all right inverses of $A$ if it has a right inverse.

**Theorem 3.7:** Let $A$ be an $m \times n$ matrix over a commutative ring $R$ with 1. The following are equivalent.

(i) A has a right inverse.

(ii) The compound matrix $C_m(A)$, which is a $1 \times \binom{n}{m}$ vector, has a right inverse.

(iii) Some linear combination of all the $m \times m$ minors of $A$ is equal to 1.

In fact, if $\sum_{\beta\in Q_{m,n}} |A_{*\beta}|c_\beta = 1$ for some elements $\{c_\beta\}_{\beta\in Q_{m,n}}$ of $R$ then, $B = (b_{ik})$ where $b_{ik} = \frac{\partial}{\partial a_{ki}}(\sum_{\beta\in Q_{m,n}} |A_{*\beta}|c_\beta)$ is a right inverse of $A$.

**Proof:** (i) $\Rightarrow$ (ii). If $A_R^{-1}$ is a right inverse of $A$, $AA_R^{-1} = I$ where I is the $m \times m$ identity matrix. Since $\rho(I) = m \leq \rho(A) \leq n$ we have that $m \leq n$. By Theorem 2.4 (b), we have that $C_m(A)C_m(A_R^{-1}) = C_m(I) = 1$. The order of $C_m(A)$ is $1 \times \binom{n}{m}$. This implies that the row vector $C_m(A)$ is a nonzero vector and has a right inverse.

(ii) $\Rightarrow$ (iii). In fact, since $C_m(A)$ has a right inverse there is a vector $\underline{y}$ of order $\binom{n}{m} \times 1$ such that $C_m(A)\underline{y} = 1$. This means that the linear combination of $m \times m$ minors of $A$ with coefficients from the vector $\underline{y}$ equals one.

(iii) $\Rightarrow$ (i). Suppose that $\sum\limits_{\beta \in Q_{m,n}} |A_{*\beta}|c_\beta = 1$ for some elements $\{c_\beta\}_{\beta \in Q_{m,n}}$ of $R$. If we write $b_{ik} = \frac{\partial}{\partial a_{ki}}(\sum\limits_\beta |A_{*\beta}|c_\beta)$, then for any fixed $k$

$$\begin{aligned}\sum_{i=1}^{n} a_{ki}b_{ik} &= \sum_\beta c_\beta \left(\sum_i a_{ki}\frac{\partial |A_{*\beta}|}{\partial a_{ki}}\right)\\ &= \sum_\beta c_\beta |A_{*\beta}| = 1.\end{aligned}$$

Now, for $\ell \neq k$, $\sum\limits_i a_{\ell i}b_{ik}$ can be expressed as $\sum\limits_\beta |D_{*\beta}|c_\beta$, where $D_{*\beta}$ stands for the $\beta$-columned submatrix of $D$, the matrix obtained from $A$ by replacing the $k$th row of $A$ with the $\ell$th row of $A$ and keeping the rest of the rows as they were. Since $|D_{*\beta}| = 0$ for all $\beta$, we have that $\sum\limits_{i=1}^{n} a_{\ell i}b_{ik} = 0$ if $\ell \neq k$. Thus, if $B$ is the $n \times m$ matrix whose $(i,k)$th element is $b_{ik}$, then $B$ is a right inverse of $A$.

♠

The previous Theorem describes a method of constructing a right inverse of $A$ from $\sum\limits_{\beta \in Q_{m,n}} |A_{*\beta}|c_\beta = 1$. In the next Theorem we shall show that every right inverse of $A$ comes from some $\sum\limits_{\beta \in Q_{m,n}} |A_{*\beta}|c_\beta = 1$.

**Theorem 3.8:** Let $A$ be an $m \times n$ matrix over a commutative ring $R$ with identity. Let $\rho(A) = m$. If $\{c_\beta\}_{\beta \in Q_{m,n}}$ are elements of $R$ such that $\sum |A_{*\beta}|c_\beta = 1$ then $B = (b_{ij})$ defined by $b_{ij} = \frac{\partial}{\partial a_{ji}}(\sum\limits_{\beta \in Q_{m,n}} |A_{*\beta}|c_\beta)$ is a right inverse of $A$. Conversely, if $C = (c_{ij})$ is a right inverse of $A$ then $\sum\limits_{\beta \in Q_{m,n}} |A_{*\beta}||C_{\beta *}| = 1$ and $c_{ij} = \frac{\partial}{\partial a_{ji}}(\sum\limits_{\beta \in Q_{m,n}} |A_{*\beta}||C_{\beta *}|)$ for all $i,j$. A similar result also holds for matrices with left inverses.

**Proof:** We have already seen in the proof of the previous Theorem that $B = (b_{ij})$ defined by $b_{ij} = \frac{\partial}{\partial a_{ji}}(\sum\limits_{\beta \in Q_{m,n}} |A_{*\beta}|c_\beta)$ is a right inverse of $A$.

Now, let $C = (c_{ij})$ be an $n \times m$ matrix such that $AC = I$.

Then by the Cauchy-Binet Theorem we get that $\sum_{\beta\in Q_{m,n}} |A_{*\beta}||C_{\beta *}| = |I| = 1$. Let us look at $\frac{\partial}{\partial a_{ji}}(\sum_{\beta\in Q_{m,n}} |A_{*\beta}||C_{\beta *}|)$ for a fixed $i$ and $j$.

$$\begin{aligned}
\frac{\partial}{\partial a_{ji}}\left(\sum_{\beta\in Q_{m,n}} |A_{*\beta}||C_{\beta *}|\right) &= \sum_{\beta\in Q_{m,n}}\left(\frac{\partial}{\partial a_{ji}}|A_{*\beta}|\sum_{\ell} c_{i\ell}\frac{\partial}{\partial c_{i\ell}}|C_{\beta *}|\right)\\
&= \sum_{\beta\in Q_{m,n}, i\in\beta}\sum_{\ell}\frac{\partial}{\partial a_{ji}}|A_{*\beta}|c_{i\ell}\frac{\partial}{\partial c_{i\ell}}|C_{\beta *}|\\
&= \sum_{\beta\in Q_{m,n}, i\in\beta}\sum_{\ell}(-1)^{j+k}|A_{\{1,2,\cdots,m\}-\{j\},\beta-\{i\}}|\\
&\quad \times c_{i\ell}(-1)^{\ell+k}|C_{\beta-\{i\},\{1,2,\cdots,m\}-\{\ell\}}|
\end{aligned}$$

where $i = \beta_k$ if $\beta$ is written as $\{\beta_1, \beta_2, \cdots, \beta_m\}$ with $\beta_1 < \beta_2 < \cdots < \beta_m$. If we call $\beta - \{i\}$ as $\gamma$ then $|\gamma| = m - 1$ and $i \notin \gamma$. Hence, the above quantity

$$= \sum_{\gamma\in Q_{m-1,n}, i\notin\gamma}\sum_{\ell}(-1)^{j+\ell}c_{i\ell}|A_{\{1,2,\cdots,m\}-\{j\},\gamma}||C_{\gamma,\{1,2,\cdots,m\}-\{\ell\}}| \qquad (*)$$

But, if $\gamma$ is such that $i \in \gamma$ then $\sum_{\ell}(-1)^{\ell}c_{i\ell}|C_{\gamma,\{1,2,\cdots,m\}-\{\ell\}}|$ is equal to

$$\begin{vmatrix} c_{i1}c_{i2}\cdots c_{im}\\ C_{\gamma,\{1,2,\cdots,m\}}\end{vmatrix}$$

Since $i \in \gamma$, $(c_{i1}\ c_{i2}\ \cdots\ c_{im})$ is a row of $C_{\gamma,\{1,2,\cdots,m\}}$. Hence this determinant is equal to zero, Thus $(*)$ can be rewritten without the restriction of $i \notin \gamma$.

Hence

$$\begin{aligned}
(*) &= \sum_{\gamma\in Q_{m-1,n}}\sum_{\ell}(-1)^{j+\ell}c_{i\ell}|A_{\{1,2,\cdots,m\}-\{j\},\gamma}||C_{\gamma,\{1,2,\cdots,m\}-\{\ell\}}|\\
&= \sum_{\ell}(-1)^{j+\ell}c_{i\ell}\sum_{\gamma\in Q_{m-1,n}}|A_{\{1,2,\cdots,m\}-\{j\},\gamma}||C_{\gamma,\{1,2,\cdots,m\}-\{\ell\}}| \qquad (**)
\end{aligned}$$

But, $\sum_{\gamma\in Q_{m-1,n}}|A_{\{1,2,\cdots,m\}-\{j\},\gamma}||C_{\gamma,\{1,2,\cdots,m\}-\{\ell\}}|$ is the $(\{1,2,\cdots,m\}-\{j\}, \{1,2,\cdots,m\}-\{\ell\})$ element of the compound matrix $C_{m-1}(AC) = I$ which is an $m\times m$ matrix. Hence

$$\sum_{\gamma\in Q_{m-1,m}}|A_{\{1,2,\cdots,m\}-\{j\},\gamma}||C_{\gamma,\{1,2,\cdots,m\}-\{\ell\}}| = 0 \text{ if } j\neq\ell \text{ and } = 1 \text{ if } j=\ell.$$

Hence,

$$(**) = (-1)^{j+j} c_{ij} = c_{ij}.$$

Thus the Theorem is proved.

♠

Similar Theorems also hold for the case of left inverse. We have already seen that over a commutative ring $R$ with 1, if an $m \times n$ matrix $A$ has a right inverse (or a left inverse) then the right inverse (the left inverse) is a $g$-inverse. Conversely, over integral domains,

**Theorem 3.9:** Let $A$ be an $m \times n$ matrix over an integral domain $R$. Let $\rho(A) = m$. Then a matrix $G$ over $R$ is a $g$-inverse of $A$ if and only if $G$ is a right inverse of $A$. A similar statement for left inverses is also valid.

**Proof:** Clearly, every right inverse is a $g$-inverse. To prove the converse let $G$ be a $g$-inverse of $A$, i.e., $AGA = A$. Since $R$ is an integral domain it can be embedded in a field and since $\rho(A) = m$, $A$ has a right inverse over the field. Let $B$ be such a right inverse over the field. Then $AG = AGAB = AB = I$. Thus, $AG = I$.

♠

Thus matrices with right inverses or left inverses are regular.

Some more matrices are also trivially regular. Let us say that a matrix $A$ over a commutative ring with 1 has a *rank factorization* if $A$ can be written as $LR$ where $L$ has a left inverse and $R$ has a right inverse. $LR$ is called a *rank factorization of* $A$. It is well known that over a field every matrix has a rank factorization. However even over integral domains every matrix need not have a rank factorization. Matrices over the ring of integers $\mathbb{Z}$ can be witness to this. Note that, if $A = LR$ is a rank factorization over a commutative ring and $A$ is of order $m \times n$ of rank $r$ then $L$ is of order $m \times r$ and $R$ is of order $r \times n$. If $A$ has a rank factorization $LR$ then, $R_R^{-1} L_L^{-1}$ where $L_L^{-1}$ is a left inverse of $L$ and $R_R^{-1}$ is a right inverse of $R$, is a $g$-inverse of $A$ since $LRR_R^{-1} L_L^{-1} LR = LR$.

Also, over a commutative ring with 1, if $A$ is idempotent and if $A = LR$ is a rank factorization then $RL = I$. Since $Tr(LR) = Tr(RL)$ we get that $\rho(A) = Tr(A)$. Thus, for any idempotent matrix with a rank factorization, rank equals trace.

We shall see later in section 4.6 that many of the usual rings have the property that every idempotent matrix has a rank factorization.

## 3.5 M-P inverses

Just as the Theorem of Von Neumann characterized all those associative rings over which every matrix is regular, in this section we shall characterize all those associative rings with identity and with an involution $a \to \bar{a}$ over which every matrix admits Moore-Penrose inverse. Surprisingly, this turns out to be a neat characterization. We shall start with some necessary and sufficient conditions for the existence of $\{1,3\}$-inverse, $\{1,4\}$-inverse and Moore-Penrose inverse. We shall give two sets of necessary and sufficient conditions for the existence of Moore-Penrose inverse and a third condition will be given in an Exercise at the end of the Chapter.

**Proposition 3.10:** Let $A$ be a matrix over an associative ring $R$ with 1 and with an involution $a \to \bar{a}$. Then

(a) $A$ has a $\{1,3\}$-inverse if and only if $A^*A$ is regular and $A$ has the property that $AC = 0$ whenever $A^*AC = 0$.

(b) $A$ has a $\{1,4\}$-inverse if and only if $AA^*$ is regular and $A$ has the property that $DA = 0$ whenever $DAA^* = 0$.

(c) $A$ has Moore-Penrose inverse if and only if $A^*A$ and $AA^*$ are both regular and $A$ has the properties that $AC = 0$ whenever $A^*AC = 0$ and $DA = 0$ whenever $DAA^* = 0$.

(d) $A$ has Moore-Penrose inverse if and only if $A^*AA^*$ is regular and $A$ has the properties that $AC = 0$ whenever $A^*AC = 0$ and $DA = 0$ whenever $DAA^* = 0$.

In case of (a), $(A^*A)^-A^*$ is a $\{1,3\}$-inverse of $A$. In case of (b), $A^*(AA^*)^-$ is a $\{1,4\}$-inverse of $A$. In case of (c), $A^*(AA^*)^-A(A^*A)^-A^*$ is the Moore-Penrose inverse of $A$. In case of (d), $A^*(A^*AA^*)^-A^*$ is the Moore-Penrose inverse of $A$.

**Proof:**

We shall prove (a) and (d). The proof of (b) is similar to that of (a). (c) follows from (a) and (b) and the comments after the Moore-Penrose equations in section 3.1.

(a) Let $G$ be a $\{1,3\}$-inverse of $A$. Then

$$A = AGAGA = AGG^*A^*A.$$

This gives us that $A^*A = A^*AGG^*A^*A$, i.e., $A^*A$ is regular. Also, $A = AGA = G^*A^*A$. This gives us that $AC = 0$ whenever $A^*AC = 0$.

Conversely, if $A^*A$ is regular, let $G = (A^*A)^-A^*$. Then $A^*AGA - A^*A = 0$. Hence $A^*A(GA - I) = 0$. Since $A$ has the property that $AC = 0$ whenever

$A^*AC = 0$ we get that $A(GA - I) = 0$, i.e., $AGA = A$. Also,

$$\begin{aligned}(AG)^* &= G^*A^* = A(A^*A)^{-*}A^* = AGA(A^*A)^{-*}A^* \\ &= A(A^*A)^- A^*A(A^*A)^{-*}A^* = A(A^*A)^-(AGA)^* \\ &= A(A^*A)^- A^* = AG.\end{aligned}$$

Thus $G$ is a $\{1,3\}$-inverse of $A$.

(d) Let $G$ be the Moore-Penrose inverse of $A$. Then $A^* = A^*G^*A^*G^*A^* = A^*G^*GAA^* = A^*G^*GAGAA^* = A^*G^*GG^*A^*AA^*$. This gives us that $G^*GG^*$ is a $g$-inverse of $A^*AA^*$. Clearly, $A$ does have the properties that $AC = 0$ whenever $A^*AC = 0$ and $DA = 0$ whenever $DAA^* = 0$.

Conversely, if $A^*AA^*$ is regular let $G = A^*(A^*AA^*)^-A^*$. Then, $A^*(AGA - A)A^* = 0$. This gives us that $A(GA - I)A^* = 0$, i.e., $(AG - I)AA^* = 0$. Hence $AGA = A$. Also,

$$\begin{aligned}(GA)^* &= A^*A(A^*AA^*)^{-*}A \\ &= A^*A(A^*AA^*)^{-*}AGA \\ &= A^*A(A^*AA^*)^{-*}AA^*(A^*AA^*)^-A^*A \\ &= (AGA)^*(A^*AA^*)^-A^*A \\ &= A^*(A^*AA^*)^-A^*A = GA.\end{aligned}$$

Similarly it can be shown that $(AG)^* = AG$ and $GAG = G$. Thus $G$ is the Moore-Penrose inverse of $A$.

♠

The above Proposition quickly gives us a characterization of all those associative rings with identity and with an involution $a \to \bar{a}$ over which every matrix admits Moore-Penrose inverse.

**Theorem 3.11:** Let $R$ be an associative ring with identity and with an involution $a \to \bar{a}$. Then the following are equivalent.

(i) Every matrix over $R$ has Moore-Penrose inverse.

(ii) Every matrix over $R$ has $\{1,3\}$-inverse.

(iii) Every matrix over $R$ has $\{1,4\}$-inverse.

(iv) Every matrix over $R$ is regular and for any matrix $A$, $A = 0$ whenever $A^*A = 0$.

(v) $R$ is a regular ring and for any $n \times 1$ vector $\underline{x}$, $\underline{x} = 0$ whenever $\underline{x}^*\underline{x} = 0$.

**Proof:**

(i) $\Rightarrow$ (ii) $\Rightarrow$ (iii) are clear. If (iii) holds, clearly every matrix over $R$ is regular. Also from Proposition 3.10 (b) we get that $DA = 0$ whenever

$DAA^* = 0$. Now, if $AA^* = 0$ let $D = AA^-$. Then, $DAA^* = 0$ and hence $A = DA = 0$. Thus (iv). (iv) $\Leftrightarrow$ (v) are clear.

(iv) $\Rightarrow$ (i) follows from Proposition 3.10 (c) or (d). This is because, if $DAA^* = 0$, it follows that $DAA^*D^* = 0$. Hence, from (iv) we get that $DA = 0$. Similarly whenever $A^*AC = 0$ we get that $AC = 0$.

♠

Thus over an associative ring with identity and with an involution, every matrix has Moore-Penrose inverse if and only if it is regular and formally real in the sense that $x_i = 0$ for all $i$ whenever $\sum_1^n \bar{x}_i x_i = 0$.

## EXERCISES:

**Exercise 3.1:** Find conditions on a matrix $A$ so that every $g$-inverse of $A$ is a reflexive $g$-inverse.

**Exercise 3.2:** (a) Over an integral domain $R$ if every matrix has a rank factorization show that $R$ is a field.

(b) Over a commutative ring $R$ if every matrix has a rank factorization show that $R$ is a field.

(c) Over a noncommutative ring $R$ with identity, if every matrix has a rank factorization is it true that $R$ is a skew field?

**Exercise 3.3:** Is it essential that $R$ has identity for the validity of the results of section 3.5?

**Exercise 3.4:** Let $R$ be an associative ring with 1 and with an involution $a \to \bar{a}$. Show that a matrix $A$ over $R$ has Moore-Penrose inverse if and only if $A^*AA^*A$ is regular and $A$ has the properties that $AC = 0$ whenever $A^*AC = 0$ and $DA = 0$ whenever $DAA^* = 0$. In this case show that $A^*A(A^*AA^*A)^-A^*$ is the Moore-Penrose inverse of $A$.

**Exercise 3.5:** Is it possible that over an associative ring every square matrix has a group inverse?

**Exercise 3.6:** Say that an associative ring $R$ is right regular if every element of $R$ is right regular. Is there a Von Neumann Theorem for right regular rings? Develop a theory of right regularity in associative rings.

# Chapter 4

# Regularity – principal ideal rings

In this Chapter we shall characterize regular matrices over principal ideal rings. First we shall look at some results on principal ideal rings.

## 4.1 Some results on principal ideal rings

Recall from Chapter 1 that a principal ideal ring $R$ is an integral domain in which every ideal is principal, i.e., if $\mathcal{I}$ is an ideal in $R$, there is an element $a \in R$ such that $\mathcal{I} = <a>$. Recall also that, for $a, b \in R$ we say that $a$ divides $b$ and write "$a|b$" if there is a $c \in R$ such that $ac = b$. Hence $a|b$ if and only if $<a> \supset <b>$. If $a$ does not divide $b$ we shall write $a \nmid b$. Of course $<a> = <b>$ if and only if there is a unit $u$ in $R$ such that $a = ub$. Sometimes we shall also write (with a slight abuse of notation) $a = b$ to mean $<a> = <b>$. If $a_1, a_2, \cdots, a_n \in R$, the ideal generated by $\{a_1, a_2, \cdots, a_n\}$ is denoted by $<a_1, a_2, \cdots, a_n>$.

In the following we shall explain the meaning of $<a_1, a_2, \cdots, a_n> = <a>$.

**Theorem 4.1:** Let $R$ be a principal ideal ring and let $a_1, a_2, \cdots, a_n$, $a$ be elements of $R$. Then the following are equivalent.

(i) $<a_1, a_2, \cdots, a_n> = <a>$.

(ii) $a|a_1, a|a_2, \cdots, a|a_n$ and $a$ is a linear combination of $a_1, a_2, \cdots, a_n$.

(iii) $a|a_1, a|a_2, \cdots, a|a_n$ and if $b \in R$ is such that $b|a_1, b|a_2, \cdots, b|a_n$ then $b|a$.

**Proof:** (i) $\Rightarrow$ (ii). From (i) we get that $<a_i> \subset <a>$ for all $i$. Hence $a|a_i$ for all $i$. Also from (i) since $a \in <a_1, a_2, \cdots, a_n>$, $a$ must equal $a_1x_1 + a_2x_2 + \cdots + a_nx_n$ for some $x_1, x_2, \cdots, x_n \in R$.

(ii) $\Rightarrow$ (iii). If $b|a_1, b|a_2, \cdots, b|a_n$, clearly $b|a_1x_1 + \cdots + a_nx_n$ for any $x_1, x_2, \cdots, x_n$ from $R$. Hence $b|a$.

(iii) $\Rightarrow$ (i). Since $R$ is a principal ideal ring there exists a $c \in R$ such that $<a_1, a_2, \cdots, a_n> = <c>$. This means that $c|a_i$ for all $i$. From (iii) we have that $c|a$. On the other hand, since $a|a_i$ for all $i$, we have that $a|c$. Thus $<c> = <a>$. This is what is required to be shown.

♠

In a principal ideal ring, if $<a_1, a_2, \cdots, a_n> = <a>$, because of (iii) above we shall call $a$ the greatest common divisor (*g.c.d.*) of $a_1, a_2, \cdots, a_n$. The *g.c.d.* is unique upto multiplication by a unit. Sometimes, we shall $<a_1, a_2, \cdots, a_n> = a$ to mean $g.c.d.(a_1, a_2, \cdots, a_n) = a$. Thus

**Theorem 4.2:** Let $a_1, a_2, \cdots, a_n$ be elements in a principal ideal ring. Then (a) the *g.c.d.* of $a_1, a_2, \cdots, a_n$ exists and it is a linear combination of $a_1, a_2, \cdots, a_n$. (b) if $a$ is a linear combination of $a_1, a_2, \cdots, a_n$ then $a$ is divisible by the *g.c.d.* of $a_1, a_2, \cdots, a_n$. (c) if 1 is a linear combination of $a_1, a_2, \cdots, a_n$ then $g.c.d.(a_1, a_2, \cdots, a_n) = 1$.

♠

A second important property of principal ideal rings is that it has no strictly increasing sequence of ideals. We shall see this now.

**Theorem 4.3:** In a principal ideal ring $R$

(a) If $\mathcal{I}_1 \subset \mathcal{I}_2 \subset \cdots$ is an infinite sequence of ideals then there is an $n$ such that $\mathcal{I}_n = \mathcal{I}_{n+1} = \cdots$.

(b) If $d_1, d_2, \cdots$ is a sequence of elements from $R$ such that $d_{i+1}|d_i$ for all $i \geq 1$ then there is an $n$ such that $d_i|d_{i+1}$ for all $i \geq n$.

**Proof:** (b) follows from (a) because "$d_{i+1}|d_i$" is equivalent to "$<d_i> \subset <d_{i+1}>$" and that "$<d_i> = <d_{i+1}>$" is equivalent to $d_i|d_{i+1}$ and $d_{i+1}|d_i$.

Let us prove (a). Let $\mathcal{I} = \bigcup_{i \geq 1} \mathcal{I}_i$. Then $\mathcal{I}$ is an ideal and hence there is an $a \in R$ such that $\mathcal{I} = <a>$. This implies that there is an $n$ such that $a \in \mathcal{I}_n$. Then clearly $\mathcal{I}_i = \mathcal{I}_{i+1} = \mathcal{I}$ for all $i \geq n$.

♠

In fact, the above properties from Theorem 4.3 characterize principal ideal rings among integral domains. Now we shall show that over a principal ideal ring there are many unimodular matrices.

**Theorem 4.4:** Let $a_1, a_2, \cdots, a_n$ be elements in a principal ideal ring $R$ such that $g.c.d.(a_1, a_2, \cdots, a_n) = d$. Then there is a matrix $A \in R_{n \times n}$ such that the first row of $A$ is $(a_1\ a_2\ \cdots\ a_n)$ and $|A| = d$. In particular, if *g.c.d.*$(a_1, a_2, \cdots, a_n) = 1$ then there is a unimodular matrix $A$ over $R$ the first row of which is $(a_1\ a_2\ \cdots\ a_n)$.

**Proof:** The second part clearly follows from the first. Let us prove the first part by induction on $n$.

For $n = 2$, from $<a_1, a_2> = d$ it follows that there are $p, q \in R$ such that $a_1p - a_2q = d$. Then $\begin{bmatrix} a_1 & a_2 \\ q & p \end{bmatrix} = A$ is a matrix with $|A| = d$.

For $n \geq 3$, let us assume the result for $n - 1$ and show that the result holds for $n$. Clearly $<a_1, a_2, \cdots, a_n> = <<a_1, a_2, \cdots, a_{n-1}>, a_n>$. Let $e = <a_1, a_2, \cdots, a_{n-1}>$ and $B \in R_{n-1 \times n-1}$ be such that its first row is $(a_1\, a_2 \cdots a_{n-1})$ and $|B| = e$. Since $<e, a_n> = d$ there exist elements $p$ and $q$ such that $ep - a_nq = d$.

Let

$$A = \begin{bmatrix} B & & & a_n \\ & & & 0 \\ & & & 0 \\ & & & \cdots \\ \frac{a_1q}{e}, \frac{a_2q}{e}, \cdots, \frac{a_{n-1}q}{e} & & & p \end{bmatrix}.$$

Since $<a_1, a_2, \cdots, a_{n-1}> = e, e|a_i$ for all $1 \leq i \leq n-1$. Thus $A$ is an $n \times n$ matrix over $R$ whose first row is $(a_1\, a_2 \cdots a_n)$. Let us evaluate $|A|$.

Expanding $|A|$ by the last column of $A$ gives us $|A| = p|B| + (-1)^{n-1}a_n|C|$ where $C$ is the submatrix of $A$ obtained by deleting the first row and the last column.

Now, since

$$|B| = a_1\frac{\partial|B|}{\partial a_1} + \cdots + a_{n-1}\frac{\partial|B|}{\partial a_{n-1}},$$

we get that

$$\frac{a_1q}{e}\frac{\partial|B|}{\partial a_1} + \cdots + \frac{a_{n-1}q}{e}\frac{\partial|B|}{\partial a_{n-1}}$$

is the determinant of the matrix $D$ obtained from $B$ by replacing the first row of $B$ by

$$\left(\frac{a_1q}{e}, \frac{a_2q}{e}, \cdots, \frac{a_{n-1}q}{e}\right).$$

Now,

$$\begin{aligned} |C| = (-1)^{n-2}|D| &= (-1)^{n-2}\frac{q}{e}|B| \\ &= (-1)^{n-2}\frac{q}{e}e \\ &= (-1)^{n-2}q. \end{aligned}$$

Hence,

$$\begin{aligned}|A| &= pe + (-1)^{n-1}(-1)^{n-2}a_n q \\ &= pe - a_n q = d.\end{aligned}$$

Thus $A$ has the properties of the required matrix.

♠

From the above Theorem it immediately follows that, whenever $c$ is a linear combination of $a_1, a_2, \cdots, a_n$ there is a matrix $C$ such that the first row of $C$ is $(a_1\, a_2\, \cdots\, a_n)$ and $|C| = c$. The following result gives the converse.

**Theorem 4.5:** Given $a_1, a_2, \cdots, a_n, c \in R$, a principal ideal ring, there is a matrix $C \in R_{n\times n}$ with its first row being $(a_1\, a_2\, \cdots\, a_n)$ and $|C| = c$ if and only if $<a_1, a_2, \cdots, a_n> \supset <c>$.

♠

## 4.2 Smith Normal Form Theorem

In matrix algebra canonical forms of matrices play an important role. For example, every matrix over a field can be written as $U\begin{bmatrix} I & 0 \\ 0 & 0 \end{bmatrix}V$ where $U$ and $V$ are nonsingular matrices. For matrices over principal ideal rings also a somewhat similar result holds and this was proved by H.F.J. Smith. We shall start with some definitions.

**Definitions:** Let $R$ be a commutative ring with 1.

(i) For matrices $A$ and $B$ over $R$ we say that $A$ and $B$ are equivalent if there are unimodular matrices $U$ and $V$ such that $B = UAV$.

(ii) For matrices $A$ and $B$ over $R$ we say that $B$ is a multiple of $A$ if there are matrices $S$ and $T$ such that $B = SAT$.

Being equivalent is clearly an equivalence relation. Also, if $A$ and $B$ are equivalent then each is a multiple of the other.

For a matrix $A$ of order $m \times n$ over a commutative ring $R$ and integer $k \geq 0$ we define $\mathcal{I}_k(A)$ as follows: $\mathcal{I}_0(A) = R$, for $1 \leq k \leq \min(m,n), \mathcal{I}_k(A)$ is the ideal generated by all $k \times k$ minors of $A$ and for other $k$'s, $\mathcal{I}_k(A) = \{0\}$. Clearly, if $\rho(A) = r > 0$ then $\mathcal{I}_0(A) \supset \mathcal{I}_1(A) \supset \cdots \supset \mathcal{I}_r(A) \neq \{0\}$ and $\mathcal{I}_{r+1}(A) = \{0\}$. If the ring is a principal ideal ring, $d_k(A)$ is defined as the *g.c.d.* of all the $k \times k$ minors of $A$, i.e., $<d_k(A)> = \mathcal{I}_k(A)$. Clearly $\mathcal{I}_k(A) = \{0\}$ for all $k > \rho(A)$ and $\mathcal{I}_k(A) \neq \{0\}$ for all $k \leq \rho(A)$.

We shall start with,

**Proposition 4.6:** Over a commutative ring if $B$ and $A$ are matrices such that $B$ is a multiple of $A$ then $\mathcal{I}_k(B) \subset \mathcal{I}_k(A)$ for all $k$. In case the ring is a principal ideal ring we also have $d_k(A)|d_k(B)$ for all $k$.

**Proof:** Since $B$ is a multiple of $A$ there exists matrices $S$ and $T$ such that $B = SAT$. Taking the $k$th compounds we get that $C_k(B) = C_k(S) C_k(A)C_k(T)$. This implies that every element of $C_k(B)$ is a linear combination of elements of $C_k(A)$. Hence $\mathcal{I}_k(B) \subset \mathcal{I}_k(A)$. In case the ring is a principal ideal ring we know that $\mathcal{I}_k(B) = <d_k(B)>$ and $\mathcal{I}_k(A) = <d_k(A)>$. Here $<d_k(B)> \subset <d_k(A)>$. This means that $d_k(A)|d_k(B)$.

♠

For equivalent matrices we have,

**Proposition 4.7:** If $A$ and $B$ are equivalent, then $\mathcal{I}_k(A) = \mathcal{I}_k(B)$ for all $k$, i.e., every $k \times k$ minor of $A$ is a linear combination of $k \times k$ minors of $B$ and every $k \times k$ minor of $B$ is a linear combination of $k \times k$ minors of $A$. In case the ring is a principal ideal ring we also have $d_k(A) = d_k(B)$ for all $k$. In particular for a $d$ in $R$ $d|a_{ij}$ for all $i, j$ if and only if $d|b_{ij}$ for all $i, j$.

**Proof:** Clearly, if $A$ and $B$ are equivalent, $A$ is a multiple of $B$ and $B$ is a multiple of $A$. The result now follows from the previous Proposition.

♠

For our Smith Normal Form Theorem we need a finer version of the previous results. This is given in the next Proposition.

**Proposition 4.8:** Let $A = BC$ and $i$ and $j$ be fixed row and column indices respectively of $A$. Let $A$ be of order $m \times n$ and $B$ be of order $m \times k$. Then the ideal

$$<a_{i1}, a_{i2}, \cdots, a_{in}> \subset <b_{i1}, b_{i2}, \cdots, b_{ik}>$$

and the ideal

$$<a_{1j}, a_{2j}, \cdots, a_{mj}> \subset <c_{1j}, c_{2j}, \cdots, c_{kj}>.$$

In case $C$ is a unimodular matrix the first inclusion will be an equality. In case $B$ is a unimodular matrix the second inclusion will be an equality.

**Proof:** Clearly $a_{il}$ for every $l$ is a linear combination of $b_{i1}, b_{i2}, \cdots, b_{ik}$. Hence the first inclusion. In case $C$ is a unimodular matrix, $AC^{-1} = B$. Hence, $<a_{i1}, a_{i2}, \cdots, a_{in}> = <b_{i1}, b_{i2}, \cdots, b_{ik}>$. The other parts are proved similarly.

♠

We shall now prove a technical result which will be instrumental in proving the main Theorem.

**Proposition 4.9:** Let $A = (a_{ij})$ be an $m \times n$ matrix with $a_{11} \neq 0$. If there exists $i_0$, $j_0$ such that $a_{11} \not| \; a_{i_0j_0}$ then $A$ is equivalent to a matrix $B = (b_{ij})$ such that $b_{11} \neq 0, b_{11}|a_{11}$ and $a_{11} \not| \, b_{11}$.

**Proof:** We shall first prove the result when $A$ is a $2 \times 2$ matrix.

Let $A = \begin{bmatrix} a_{11} & a_{12} \\ a_{21} & a_{22} \end{bmatrix}$. Let us look at the case $a_{11} \not| \; a_{12}$. Let $d =$ $g.c.d.(a_{11}, a_{12})$. Then there exists $x, y \in R$ such that $xa_{11} + ya_{12} = d$. Then $g.c.d.(x, y) = 1$. Using Theorem 4.4 (or otherwise), let $U = \begin{bmatrix} x & a \\ y & b \end{bmatrix}$ be a unimodular matrix. Then $AU$ is a matrix whose $(1,1)$ element is $d$. Also $d|a_{11}$ and $a_{11} \not| \; d$ since $a_{11} \not| \; a_{12}$ and $d|a_{12}$. Thus $B = AU$ satisfies the requirements of the Theorem.

The case of $a_{11} \not| \, a_{21}$ can also be treated in a similar fashion.

So now, let $a_{11}|a_{12}, a_{11}|a_{21}$ and $a_{11} \not| \, a_{22}$. Let us do a few multiplications with unimodular matrices.

$$\begin{bmatrix} a_{11} & a_{12} \\ a_{21} & a_{22} \end{bmatrix} \begin{bmatrix} 1 & \frac{-a_{12}}{a_{11}} \\ 0 & 1 \end{bmatrix} = \begin{bmatrix} a_{11} & 0 \\ a_{21} & \frac{-a_{21}a_{12}}{a_{11}} + a_{22} \end{bmatrix}$$

$$\begin{bmatrix} 1 & 0 \\ \frac{-a_{21}}{a_{11}} & 1 \end{bmatrix} \begin{bmatrix} a_{11} & 0 \\ a_{21} & \frac{-a_{21}a_{12}}{a_{11}} + a_{22} \end{bmatrix} = \begin{bmatrix} a_{11} & 0 \\ 0 & \frac{-a_{21}a_{12}}{a_{11}} + a_{22} \end{bmatrix}$$

$$= \begin{bmatrix} a_{11} & 0 \\ 0 & c \end{bmatrix} \text{ say.}$$

Of course $A$ is equivalent to $\begin{bmatrix} a_{11} & 0 \\ 0 & c \end{bmatrix}$ and $a_{11} \not| \, c$ since $a_{11} \not| \, a_{22}, a_{11}|a_{12}$ and $a_{11}|a_{21}$, by the last part of the statement of Proposition 4.7.

The matrix $C = \begin{bmatrix} a_{11} & c \\ 0 & c \end{bmatrix} = \begin{bmatrix} 1 & 1 \\ 0 & 1 \end{bmatrix} \begin{bmatrix} a_{11} & 0 \\ 0 & c \end{bmatrix}$ is equivalent to $A$ and $a_{11} \not| \; c$ which is the $(1,2)$-element of $C$. From the first paragraph of this proof, $C$ is equivalent to a matrix $B = \begin{bmatrix} b_{11} & b_{12} \\ b_{21} & b_{22} \end{bmatrix}$ where $b_{11}|a_{11}$ but $a_{11} \not| \, b_{11}$. Thus the given matrix $A$ is equivalent to $B$ and $b_{11}|a_{11}$ but $a_{11} \not| \, b_{11}$. Hence the proof is complete for the case of $2 \times 2$ matrices.

In the general case of $A = (a_{ij})$ being an $m \times n$ matrix with $a_{11} \neq 0$ and $i_0, j_0$ such that $a_{11} \not| \, a_{i_0j_0}$ we proceed as follows.

If $a_{11} \not| \; a_{1j_0}$ interchange the second column of $A$ with the $j_0$th column and call this matrix $C$.

If $a_{11} \not| \, a_{i_01}$, interchange the second row of $A$ with the $i_0$th row and call this matrix $C$.

If $a_{11}|a_{1j_0}$ and $a_{11}|a_{i_01}$, (but $a_{11} \not| \, a_{i_0j_0}$) interchange the second column of $A$ with the $j_0$th column and the second row of $A$ with the $i_0$th row and call this matrix $C$. Let $C = (c_{ij})$. Clearly $c_{11} = a_{11}$ and either $c_{11} \not| \, c_{12}$ or $c_{11} \not| \, c_{21}$ or $c_{11}|c_{12}$, $c_{11}|c_{21}$ and $c_{11} \not| \, c_{22}$.

Partition the matrix $C$ as $\begin{bmatrix} C_1 & C_2 \\ C_3 & C_4 \end{bmatrix}$ where $C_1$ is a $2 \times 2$ matrix. From the first part of this proof for $2 \times 2$ matrices, let $U_1$ and $V_1$ be $2 \times 2$ unimodular matrices such that $U_1C_1V_1 = B_1$ is a matrix such that the $(1,1)$ element of $B_1$ divides the $(1,1)$ element of $C_1$ but the $(1,1)$ element of $C_1$ does not divide the $(1,1)$ element of $B_1$. Then the $m \times m$ matrix $U = \begin{bmatrix} U_1 & 0 \\ 0 & I \end{bmatrix}$ and the $n \times n$ matrix $V = \begin{bmatrix} V_1 & 0 \\ 0 & I \end{bmatrix}$ both are unimodular matrices and $UCV$ will look like $\begin{bmatrix} B_1 & * \\ * & * \end{bmatrix} = B$, say. Now, the matrix $B$ is equivalent to $A$ and since $c_{11} = a_{11}$ we get that the $(1,1)$ element of $B$ divides $a_{11}$ but $a_{11}$ does not divide the $(1,1)$ element of $B$. Hence the Proposition.

♠

Now, we have all the machinery required for proving the Smith Normal Form Theorem.

**Theorem 4.10 (Smith Normal Form Theorem):** Let $A$ be an $m \times n$ matrix of rank $r$ over a principal ideal ring $R$. Then there exist unimodular matrices $U$ of order $m \times m$ and $V$ of order $n \times n$ such that $UAV$ is a diagonal matrix

$$\begin{bmatrix} s_1 & 0 & \cdots & \cdots & 0 \\ 0 & s_2 & \cdots & \cdots & 0 \\ \cdots & \cdots & \cdots & \cdots & \cdots \\ 0 & \cdots & s_r & \cdots & 0 \\ \cdots & \cdots & \cdots & \cdots & \cdots \\ 0 & 0 & \cdots & \cdots & 0 \end{bmatrix}$$

where $s_1, s_2, \cdots, s_r$ are all nonzero and $s_1|s_2, s_2|s_3, \cdots, s_{r-1}|s_r$.

Also such a respresentation is unique in the sense that if $U^0$ and $V^0$ are unimodular matrices such that $U^0AV^0$ is a diagonal matrix

$$\begin{bmatrix} t_1 & 0 & \cdots & \cdots & 0 \\ 0 & t_2 & \cdots & \cdots & 0 \\ \cdots & \cdots & \cdots & \cdots & \cdots \\ 0 & 0 & \cdots & t_r & 0 \\ \cdots & \cdots & \cdots & \cdots & \cdots \\ 0 & 0 & \cdots & \cdots & 0 \end{bmatrix}$$

where $t_1, t_2, \cdots, t_r$ are all nonzero and $t_1|t_2, t_2|t_3, \cdots, t_{r-1}|t_r$ then $s_1 = t_1$, $s_2 = t_2, \cdots, s_r = t_r$.

Further for $1 \leq k \leq r$ the $s_k$'s can be identified by

$$s_k = \frac{g.c.d. \text{ of all the } k \times k \text{ minors of } A}{g.c.d. \text{ of all the } (k-1) \times (k-1) \text{ minors of } A} = \frac{d_k(A)}{d_{k-1}(A)}.$$

**Proof:** We shall prove the first part of the Theorem by induction on $r$, the rank of $A$.

If $\rho(A) = 0$ then $A = 0$ and there is nothing to be proved. Let us now assume that the result is true for all matrices with rank $r-1$.

If $\rho(A) = r > 0$ then $A$ is a nonzero matrix. By permuting the rows and columns, if necessary, we can make sure that the $(1,1)$ element of $A$ is nonzero. So let $A_1 = A = (a_{ij})$ with $a_{11} \neq 0$. Let $f_1 = a_{11}$. If there is an $(i,j)$ such that $f_1 \nmid a_{ij}$ by Proposition 4.9 there exists unimodular matrices $U_1$ and $V_1$ such that for the matrix $U_1 A_1 V_1 = (A_2$, say), if the $(1,1)$ element of $A_2$ is $f_2$ then $f_2 \neq 0, f_2|f_1$, and $f_1 \nmid f_2$. If $f_2$ divides every element of $A_2$ then we write $B = A_2$. If there is an $(i,j)$ such that $f_2$ does not divide the $(i,j)$th element of $A_2$ then applying Proposition 4.9 to $A_2$ gives unimodular matrices $U_2$ and $V_2$ such that for the matrix $U_2 A_2 V_2 = A_3$, (say), if the $(1,1)$ element of $A_3$ is $f_3$ then $f_3 \neq 0$, $f_3|f_2$ but $f_2 \nmid f_3$.

Proceeding in this way, if for every $k \geq 1$, for the matrix $A_k$, if we call the $(1,1)$ element as $f_k$, if there is an $(i,j)$ such that $f_k$ does not divide the $(i,j)$th element of $A_k$, then we would be constructing elements $f_1, f_2, \cdots$, an infinite sequence of nonzero elements of $R$ such that $f_{i+1}|f_i$ and $f_i \nmid f_{i+1}$ for all $i \geq 1$. But since $R$ is a principal ideal ring, by Theorem 4.3 (b), this is impossible.

Hence there is a $k \geq 1$ such that for the matrix $A_k$ if $f_k$ is the $(1,1)$ element of $A_k$ then $f_k \neq 0$ and $f_k$ divides every element of $A_k$. Thus, $A$ is equivalent to a matrix $A_k = B = (b_{ij})$ say, such that $b_{11} \neq 0$ and $b_{11}|b_{ij}$ for all $i, j$.

By using elementary operations of type (ii), i.e., addition of a multiple of a particular row (or column) to another row (or column), we see that $B$ is equivalent to a matrix $C = (c_{ij})$ such that $c_{11} = b_{11} \neq 0, c_{1j} = 0$ for all $j \geq 1$ and $C_{i1} = 0$ for all $i \geq 1$.

For example, if $c_{12} \neq 0$, multiplying the first column of $C$ by $\frac{c_{12}}{c_{11}}$ and subtracting from the second column will result in a matrix whose $(1,1)$ element is $c_{11}$ and $(1,2)$ element is 0.

Hence $C$ is of the form $\begin{bmatrix} c_{11} & 0 \\ 0 & C_1 \end{bmatrix}$.

Since $c_{11} \neq 0$ and since $\rho(C) = \rho(C_1) + 1$, we see that $\rho(C_1) = r - 1$. Hence by the induction hypothesis there exist unimodular matrices $U^*$ and $V^*$ such that

$$U^* C_1 V^* = \begin{bmatrix} s_2 & 0 & \cdots & 0 \\ 0 & s_3 & \cdots & 0 \\ \cdots & \cdots & \cdots & \cdots \\ 0 & 0 & s_r & 0 \\ \cdots & \cdots & \cdots & \cdots \\ 0 & 0 & \cdots & 0 \end{bmatrix} (= D \text{ say}).$$

where $s_2|s_3, s_3|s_4, \cdots, s_{r-1}|s_r$.

Let $c_{11} = s_1$. Since $c_{11}|c_{ij}$ for all $i, j$ and since $C$ looks like $\begin{bmatrix} c_{11} & 0 \\ 0 & C_1 \end{bmatrix}$, by the last part of Proposition 4.7 we have that $c_{11}$ divides every element of $C_1$. But $C_1$ being equivalent to $D$ we have that $s_1 = c_{11}$ divides $s_2$. Also

$$\begin{bmatrix} 1 & 0 \\ 0 & U^* \end{bmatrix} C \begin{bmatrix} 1 & 0 \\ 0 & V^* \end{bmatrix} = \begin{bmatrix} s_1 & 0 & \cdots & 0 \\ 0 & s_2 & \cdots & 0 \\ \cdots & \cdots & \cdots & \cdots \\ 0 & 0 & s_r & 0 \\ \cdots & \cdots & \cdots & \cdots \\ 0 & 0 & \cdots & 0 \end{bmatrix}$$

and $\begin{bmatrix} 1 & 0 \\ 0 & U^* \end{bmatrix}$ and $\begin{bmatrix} 1 & 0 \\ 0 & V^* \end{bmatrix}$ are unimodular matrices. Thus $A$ is equivalent to

$$\begin{bmatrix} s_1 & 0 & \cdots & 0 \\ 0 & s_2 & \cdots & 0 \\ \cdots & \cdots & \cdots & \cdots \\ 0 & 0 & s_r & 0 \\ \cdots & \cdots & \cdots & \cdots \\ 0 & 0 & \cdots & 0 \end{bmatrix}$$

and $s_1|s_2, s_2|s_3, \cdots s_{r-1}|s_r$.

Thus the first part of Smith Normal Form Theorem is proved.

For the second part, let

$$S = \begin{bmatrix} s_1 & 0 & \cdots & 0 \\ 0 & s_2 & \cdots & 0 \\ \cdots & \cdots & \cdots & \cdots \\ 0 & 0 & s_r & 0 \\ \cdots & \cdots & \cdots & \cdots \\ 0 & 0 & \cdots & 0 \end{bmatrix} \text{ and } E = \begin{bmatrix} t_1 & 0 & \cdots & 0 \\ 0 & t_2 & \cdots & 0 \\ \cdots & \cdots & \cdots & \cdots \\ 0 & 0 & t_r & 0 \\ \cdots & \cdots & \cdots & \cdots \\ 0 & 0 & \cdots & 0 \end{bmatrix}$$

be equivalent matrices such that $s_i \neq 0$ for all $i, t_i \neq 0$ for all $i, s_i | s_{i+1}$ for all $1 \leq i \leq r-1$ and $t_i | t_{i+1}$ for all $1 \leq i \leq r-1$.

The *g.c.d.* of all the $1 \times 1$ minors of $D = d_1(D) = s_1$ and the *g.c.d.* of all the $1 \times 1$ minors of $E = d_1(E) = t_1$.

By Proposition 4.7, $s_1 = t_1$.

$d_2(D)$, the *g.c.d.* of all the $2 \times 2$ minors of $D$ is clearly seen to be, because $s_1|s_2, s_2|s_3, \cdots, s_{r-1}|s_r$, equal to $s_1 s_2$. Similarly $d_2(E) = t_1 t_2$. Hence, by Proposition 4.7, $s_1 s_2 = t_1 t_2$. But since $s_1 = t_1$ we have that $s_2 = t_2$. Continuing in this way we get that $s_i = t_i$ for all $1 \leq i \leq r$.

Since $d_k(A) = d_k(D) = s_1 s_2 \cdots s_k$ for all $1 \leq k \leq r$ it follows that $s_k = \frac{d_k(A)}{d_{k-1}(A)}$. Thus we get the last part of the Theorem.

♠

$s_1, s_2, \cdots, s_r$ of a matrix $A$ given above are called the *invariant factors* of the matrix $A$.

Thus we have not only shown that every matrix over a principal ideal ring is equivalent to a diagonal matrix, but also that we have identified the diagonal matrix and showed its uniqueness.

## 4.3 Regular matrices over Principal Ideal Rings

In this section we shall characterize regular matrices over principal ideal rings. We shall first look at diagonal matrices. Recall that a principal ideal ring is an integral domain with 1 in which every ideal is principal.

**Proposition 4.11:** Over a principal ideal ring, more generally, over an integral domain, a matrix $A = \begin{bmatrix} S & 0 \\ 0 & 0 \end{bmatrix}$ where $S = $ *diag* $(s_1, s_2, \cdots, s_r)$ with $s_i \neq 0$ for $1 \leq i \leq r$ is regular if and only if $s_1, s_2, \cdots, s_r$ are all units in $R$. Further, a matrix $G$ is a $g$-inverse of $A$ if and only if it is of the form $\begin{bmatrix} S^{-1} & B \\ C & D \end{bmatrix}$ where $S^{-1} = $ *diag* $(s_1^{-1}, s_2^{-1}, \cdots, s_r^{-1})$.

**Proof:** If $G = \begin{bmatrix} T & B \\ C & D \end{bmatrix}$ is a $g$-inverse of $A$, using the equation $AGA = A$ we get that $STS = S$. If $T = (t_{ij})$ then $STS = S$ gives us that $s_i t_{ii} s_i = s_i$ and $s_i t_{ij} s_j = 0$ for all $i \neq j$. Since $s_i \neq 0$ for all $i$, we get that $t_{ij} = 0$ for all $i \neq j$ and $t_{ii} = s_i^{-1}$ for all $1 \leq i \leq r$.

Thus the diagonal matrix *diag* $(s_1^{-1}, s_2^{-1}, \cdots, s_r^{-1}) = S^{-1} = T$. Thus every $g$-inverse of $A$ is of the form $\begin{bmatrix} S^{-1} & B \\ C & D \end{bmatrix}$.

Now, we can characterize regular matrices over principal ideal rings.

**Theorem 4.12:** Over a principal ideal ring $R$, an $m \times n$ matrix $A$ of rank $r$ is regular if and only if

$$A = U \begin{bmatrix} I & 0 \\ 0 & 0 \end{bmatrix} V$$

where $U$ and $V$ are some unimodular matrices and $I$ is the $r \times r$ identity matrix . Further an $n \times m$ matrix $G$ is a $g$-inverse of $A$ if and only if it can be written in the form

$$V^{-1} \begin{bmatrix} I & B \\ C & D \end{bmatrix} U^{-1}$$

for some matrices $B, C$ and $D$.

**Proof:** Let $A = U \begin{bmatrix} S & 0 \\ 0 & 0 \end{bmatrix} W$ where $U$ and $W$ are unimodular matrices and $S = diag\ (s_1, s_2, \cdots, s_r)$ where $s_i \neq 0$ for all $i$ and $s_i | s_{i+1}$ for $i = 1, 2, \cdots, r-1$, be the representation of $A$ given by the Smith Normal Form Theorem. If $A$ is regular let $G$ be a $g$-inverse of $A$. Then,

$$U \begin{bmatrix} S & 0 \\ 0 & 0 \end{bmatrix} WGU \begin{bmatrix} S & 0 \\ 0 & 0 \end{bmatrix} W = U \begin{bmatrix} S & 0 \\ 0 & 0 \end{bmatrix} W.$$

Since $U$ and $W$ are unimodular matrices, $WGU$ is a $g$-inverse of $\begin{bmatrix} S & 0 \\ 0 & 0 \end{bmatrix}$. Thus $\begin{bmatrix} S & 0 \\ 0 & 0 \end{bmatrix}$ is regular. By Proposition 4.11 $s_1, s_2, \cdots, s_r$ are all units. Now

$$A = U \begin{bmatrix} S & 0 \\ 0 & 0 \end{bmatrix} W = U \begin{bmatrix} I & 0 \\ 0 & 0 \end{bmatrix} \begin{bmatrix} S & 0 \\ 0 & I \end{bmatrix} W$$

and $\begin{bmatrix} S & 0 \\ 0 & I \end{bmatrix}$ is a unimodular matrix. Hence $\begin{bmatrix} S & 0 \\ 0 & I \end{bmatrix} W$ is also a unimodular matrix and $A$ can be written as $U \begin{bmatrix} I & 0 \\ 0 & 0 \end{bmatrix} V$ Since every $g$-inverse of $\begin{bmatrix} I & 0 \\ 0 & 0 \end{bmatrix}$ can be written as $\begin{bmatrix} I & B \\ C & D \end{bmatrix}$ we have that every $g$-inverse of $U \begin{bmatrix} I & 0 \\ 0 & 0 \end{bmatrix} V$ can be written as $V^{-1} \begin{bmatrix} I & B \\ C & D \end{bmatrix} U^{-1}$ for some $B, C$ and $D$.

♠

Now we shall give a more amenable Theorem to characterize regular matrices over principal ideal rings.

**Theorem 4.13:** Let $A$ be an $m \times n$ matrix of rank $r$ over a principal ideal ring. Then the following are equivalent.

(i) $A$ is regular.

(ii) $A = U \begin{bmatrix} I & 0 \\ 0 & 0 \end{bmatrix} V$ for some unimodular matrices $U$ and $V$.

(iii) $g.c.d.$ of all the $r \times r$ minors of $A$ is 1.

(iv) a linear combination of all the $r \times r$ minors of $A$ is 1.

(v) $A$ has a rank factorization.

**Proof:** (i) $\Leftrightarrow$ (ii) was proved in Theorem 4.12. From this it also follows that $A$ is equivalent to $\begin{bmatrix} I & 0 \\ 0 & 0 \end{bmatrix}$. But there is only one $r \times r$ nonzero minor of $\begin{bmatrix} I & 0 \\ 0 & 0 \end{bmatrix}$ and it is equal to $|I| = 1$. Hence by Proposition 4.7 we have that the $g.c.d.$ of all the $r \times r$ minors of $A$ is 1. Thus (ii) $\Rightarrow$ (iii) is proved. By Theorem 4.2 (a), (iii) $\Rightarrow$ (iv) is clear. By Theorem 4.2 (c), (iv) $\Rightarrow$ (iii) is clear.

If (iii) holds we get that $d_r(A) = 1$. But $d_r(A) = s_1 s_2 \cdots s_n$ as in the Smith Normal form of $A$. Hence $s_1, s_2, \cdots, s_n$ are all units in $R$. The last part of Theorem 4.10 completes the proof that (iii) $\Rightarrow$ (ii).

Now let (ii) hold. If we partition $U$ as $[\; U_1 \quad U_2 \;]$ and $V$ as $\begin{bmatrix} V_1 \\ V_2 \end{bmatrix}$ to conform to the product $A = [\; U_1 \quad U_2 \;] \begin{bmatrix} I & 0 \\ 0 & 0 \end{bmatrix} \begin{bmatrix} V_1 \\ V_2 \end{bmatrix}$ then we get $A = U_1 V_1$. Since $U$ and $V$ are units $U_1$ has a left inverse and $V_1$ has a right inverse. Thus $A$ has a rank factorization. If on the other hand, if $A$ has a rank factorization then we have already seen in the discussion after Theorem 3.9 that $A$ is regular. This proves (ii) $\Rightarrow$ (v) $\Rightarrow$ (i).

♠

In case $A$ is regular, let us find all its $g$-inverses using the characterization (ii) of the above Theorem.

**Theorem 4.14:** Over any principal ideal ring (in fact over any associative ring) if $A = U \begin{bmatrix} I & 0 \\ 0 & 0 \end{bmatrix} V$ for some unimodular matrices $U$ and $V$ then all $g$-inverses of $A$ are of the form $V^{-1} \begin{bmatrix} I & B \\ C & D \end{bmatrix} U^{-1}$ for some matrices $B, C$ and $D$ of appropriate order.

**Proof:** Clearly follows from the fact that all $g$-inverse of $\begin{bmatrix} I & 0 \\ 0 & 0 \end{bmatrix}$ are given by matrices of the form $\begin{bmatrix} I & B \\ C & D \end{bmatrix}$.

♠

## 4.4 An algorithm for Euclidean domains

Given a matrix $A$ over the ring $\mathbb{Z}$ how do we verify whether $A$ is regular or not? If $A$ is regular how do we find a $g$-inverse of $A$?

In this section we shall give an algorithm which answers these questions for matrices over any Euclidean domain. Thus, this algorithm is applicable for matrices over the ring of polynomials in one variable with real coefficients (or coefficients from any field) also.

We need a result which holds good even over integral domains.

**Theorem 4.15:** For matrices over integral domains,

(a) If a matrix $\begin{bmatrix} a & \underline{z}^T \\ \underline{0} & A \end{bmatrix}$ where $a \neq 0, \underline{z}^T = (z_1, z_2, \cdots, z_n)$ and $A$ is an $m \times n$ matrix, has a $g$-inverse then some linear combination of $a, z_1, z_2, \cdots, z_n$ equals 1. In case the integral domain is a principal ideal ring, then, the $g.c.d.(a, z_1, z_2, \cdots, z_n) = 1$.

(b) If a matrix $\begin{bmatrix} a & \underline{0}^T \\ \underline{0} & A \end{bmatrix}$ where $A$ is an $m \times n$ matrix, has a $g$-inverse then either $a = 0$ or a unit and $A$ has a $g$-inverse.

**Proof:** Let $\begin{bmatrix} b & \underline{x}^T \\ \underline{y} & B \end{bmatrix}$ be a $g$-inverse of $\begin{bmatrix} a & \underline{z}^T \\ \underline{0} & A \end{bmatrix}$. Then

$$\begin{bmatrix} a & \underline{z}^T \\ \underline{0} & A \end{bmatrix}\begin{bmatrix} b & \underline{x}^T \\ \underline{y} & B \end{bmatrix}\begin{bmatrix} a & \underline{z}^T \\ \underline{0} & A \end{bmatrix} = \begin{bmatrix} a & \underline{z}^T \\ \underline{0} & A \end{bmatrix}.$$

Equating the $(1,1)$ elements of both sides gives us

$$aba + \underline{z}^T \underline{y} a = a$$

Since the ring has no zero divisors and since $a \neq 0$ we get that $ab + \underline{z}^T\underline{y} = 1$. Hence, some linear combination of $a, z_1, z_2, \cdots, z_n$ equals 1. In case the ring is a principal ideal ring, Theorem 4.2 (c) gives us that $g.c.d.(a, z_1, z_2, \cdots, z_n) = 1$. This proves (a).

To prove (b), let $\begin{bmatrix} b & \underline{x}^T \\ \underline{y} & B \end{bmatrix}$ be a $g$-inverse of $\begin{bmatrix} a & \underline{0}^T \\ \underline{0} & A \end{bmatrix}$.

Then,

$$\begin{bmatrix} a & \underline{0}^T \\ \underline{0} & A \end{bmatrix}\begin{bmatrix} b & \underline{x}^T \\ \underline{y} & B \end{bmatrix}\begin{bmatrix} a & \underline{0}^T \\ \underline{0} & A \end{bmatrix} = \begin{bmatrix} a & \underline{0}^T \\ \underline{0} & A \end{bmatrix}.$$

i.e.,

$$\begin{bmatrix} aba & a\underline{x}^T A \\ A\underline{y}a & ABA \end{bmatrix} = \begin{bmatrix} a & \underline{0}^T \\ \underline{0} & A \end{bmatrix}.$$

Equating the parts gives us that $aba = a$ and $ABA = A$. Thus, $A$ is regular. Also $(ab-1)a = 0$. Since our ring is an integral domain we get that $a = 0$ or $ab = 1$ Thus $a = 0$ or is a unit.

♠

The previous Theorem has the following consequence for matrices over principal ideal rings. Let $A$ be a nonzero matrix with the first column being nonzero. By taking a linear combination of the elements of the first column replace the $(1,1)$ element by the *g.c.d.* of elements of the first column. This can be achieved by premultiplying the given matrix by a unimodular matrix. Now, using the $(1,1)$ element make all the other elements of the first column zero. This can be achieved by a further premultiplication by unimodular matrices corresponding to elementary operations. Call the new matrix $B$. If $A$ is regular, the previous Theorem tells us that the *g.c.d.* of the elements of the first row of $B$ must be a unit.

We shall use the above observation in the algorithm below for Euclidean domains. Let $(R, |\cdot|)$ be a Euclidean domain. The Euclidean algorithm says that, for any $a$ and $b$ in $R$, where $a \neq 0$, there exists $c$ and $d$ in $R$ such that $b = ac + d$ where $d = 0$ or $|d| < |a|$. Let us see how the Euclidean algorithm can be used to find the *g.c.d.* of two nonzero elements of $R$. Call $a = a_1$ and $b = b_1$. If $|b_1| < |a_1|$ take $d = b_1$ and $c = 0$. If $|a_1| \leq |b_1|$, write $b_1 = a_1c_1 + b_2$ where $b_2 = 0$ or $|b_2| < |a_1|$. If $b_2 = 0$ then $g.c.d.(a,b) = a$. If $b_2 \neq 0$, because $b_1 = a_1c_1 + b_2$ it follows that $g.c.d.(a_1, b_1) = g.c.d.(a_1, b_2)$. Thus $b_2 = b_1 - a_1c_1$, i.e., a multiple of $a_1$ (namely, $a_1c_1$) is subtracted from $b_1$ to obtain $b_2$ with either $b_2$ being 0 or $0 \neq |b_2| < |a_1|$. Again, by the Euclidean algorithm write $a_1 = b_2c_2 + a_2$ where $a_2 = 0$ or $|a_2| < |b_2|$. If $b_2 = 0$, $g.c.d.(a,b) = g.c.d.(a_1,b_2) = g.c.d.(a_2,b_2)$. Proceeding in this way, since the Euclidean norm is a nonnegative integer valued function, it follows that this process will terminate in finitely many steps and we will obtain $g.c.d.(a,b)$.

The above principle can be generalized to find the *g.c.d.* of any finite number of elements. It is exactly this principle and the observation in the previous two paragraphs which make the following algorithm work.

**Algorithm 4.16:** Let $(R, |\cdot|)$ be a Euclidean ring. We shall first describe two operations on triples of matrices. Let $D, E, F$ be $m \times m, m \times n$ and $n \times n$ matrices over $R$. Let $E = (e_{ij})$. $R_p(D,E,F)$ for $1 \leq p \leq k = \min(m,n)$ is defined as follows:

Consider the $(p,p), (p+1,p), \cdots, (m,p)$ elements of $E$. If all of them are zero we set $R_p(D,E,F) = (D,E,F)$. If at least one of $e_{pp}, e_{p+1p}, \cdots, e_{mp}$ is $\neq 0$, among the nonzero elements of $e_{pp}, e_{p+1p}, \cdots, e_{mp}$ look for the one with minimal $|\cdot|$. If there are many, choose that $e_{ip}$ such that $i$ is the minimum (i.e.,

$|e_{ip}| \leq |e_{jp}|$ whenever $|e_{jp}| \neq 0$ and $p \leq j \leq m$ and $i$ is the smallest such). Let $e_{ip} \neq 0$ be such that $p \leq i \leq m$ and such that $|e_{ip}| \leq |e_{jp}|$ for every $j$ such that $p \leq j \leq m$ and $e_{jp} \neq 0$. Now subtract multiples of $i$th row of $E$ from the $j$th row of $E$ (for $j$ such that $p \leq j \leq m$ and $e_{jp} \neq 0$ and $j \neq i$) so that the new $(j,p)$ element is either equal to zero or has less $|\cdot|$ than $|e_{ip}|$. Repeat this process on the new matrix and the $p, p+1, \cdots, m$th rows of $E$ many times until we get a situation where only one of the $(p,p), (p+1,p), \cdots, (m,p)$ elements has nonzero $|\cdot|$. This element with nonzero $|\cdot|$ is exactly a *g.c.d.* of $e_{pp}, e_{p+1p}, \cdots, e_{mp}$. If this is the $(i_0, p)$ element, interchange the $p$th and the $i_0$th rows of $E$. For this matrix the $(p,p)$ element is nonzero. If this $(p,p)$ element is a unit (say $u$), multiply the $p$th row of this matrix by $u^{-1}$ and call the matrix $E^*$. If the $(p,p)$ element is not a unit call this matrix itself $E^*$. For the matrix $E^*$, the $(p,p)$ element is a *g.c.d.* of $(p,p), (p+1,p), \cdots, (m,p)$ elements of $E$. Also the $(p+1,p), \cdots, (m,p)$ elements of $E^*$ are all zero and the $(p,p)$ element is either 1 or is not a unit.

Now, observe that all the operations done above are row operations. Perform the whole sequence of row operations that are performed on $E$, on $D$ also and call the new matrix $D^*$. We define $R_p(D,E,F) = (D^*, E^*, F)$.

We define $C_p(D,E,F)$ for $1 \leq p < k$ in a similar fashion starting with $(p,p), (p,p+1), \cdots, (p,m)$ elements of $E$ and performing the necessary column operations. This time all the sequence of column operations that are done on $E$ are also performed on $F$.

Now we are ready to give our algorithm.

Let $A$ be an $m \times n$ matrix for which we want to decide whether it is regular and produce a $g$-inverse if it is regular. Let $k = min(m,n)$.

Step 1. Let $p = 1$, $D_1 = I$ the $m \times m$ identity matrix, $F_1 = I$ the $n \times n$ identity matrix and $A_1 = A$. Go to Step 2.

Step 2. Find $C_p(R_p(C_p(R_p(D_p, A_p, F_p))))$ and call it $(D_{p+1}, A_{p+1}, F_{p+1})$. If the $(p,p)$ element of $A_{p+1}$ is neither 0 nor 1 conclude that $A$ is not regular and stop. If $p = k$ and the $(p,p)$ element of $A_{p+1}$ is either 0 or 1 go to Step 3. If $p < k$ and the $(p,p)$ element of $A_{p+1}$ is either 0 or 1 increase the value of $p$ by 1 and go back to Step 2.

Step 3. Compute the product $F_{k+1}A^T_{k+1}D_{k+1}$ and stop.

Now we have

**Theorem 4.17:** (a) If for some $1 \leq p \leq k$ the $(p,p)$ element of $A_{p+1}$ is neither 0 nor 1 then $A$ is not regular.

(b) The above algorithm proceeds upto $p = k$ and the $(k,k)$ element of $A_{k+1}$ is either 0 or 1 if and only if $A$ is regular. In this case $F_{k+1}A^T_{k+1}D_{k+1}$ is a $g$-inverse of $A$.

**Proof:** (a) follows from Theorem 4.16.

(b) Using the results of the last two paragraphs of section 2.1 one can see that $D_2AF_2 = D_2A_1F_2 = A_2, D_3AF_3 = A_3, \cdots, D_{k+1}AF_{k+1} = A_{k+1}$ and that all the $D_i$'s and $F_i$'s are unimodular. If the $(k, k)$ element of $A_{k+1}$ is 0 or 1 then $A_{k+1}$ will be an $m \times n$ diagonal matrix with diagonal elements being 0 or 1. Hence $A_{k+1}$ is regular and $A_{k+1}A_{k+1}^T A_{k+1} = A_{k+1}$; i.e., $A_{k+1}^T$ is a $g$-inverse of $A_{k+1}$. Thus

$$D_{k+1}AF_{k+1}A_{k+1}^T D_{k+1}AF_{k+1} = D_{k+1}AF_{k+1}.$$

Since $D_{k+1}$ and $F_{k+1}$ are unimodular matrices we get that

$$AF_{k+1}A_{k+1}^T D_{k+1}A = A.$$

Thus, $F_{k+1}A_{k+1}^T D_{k+1}$ is a $g$-inverse of $A$.

♠

Let us illustrate the algorithm by an example.

**Illustration:** Let $A = \begin{bmatrix} 0 & 3 & 0 \\ 0 & 1 & 1 \end{bmatrix}$

Then,

$$\begin{aligned}
(D_1, A_1, F_1) &= \left( \begin{matrix} 1 & 0 \\ 0 & 1 \\ \ \end{matrix} \quad \begin{matrix} 0 & 3 & 0 \\ 0 & 1 & 1 \\ \ \end{matrix} \quad \begin{matrix} 1 & 0 & 0 \\ 0 & 1 & 0 \\ 0 & 0 & 1 \end{matrix} \right) \\
R_1(D_1, A_1, F_1) &= (D_1, A_1, F_1) \\
C_1(R_1(D_1, A_1, F_1)) &= \left( \begin{matrix} 1 & 0 \\ 0 & 1 \\ \ \end{matrix} \quad \begin{matrix} 3 & 0 & 0 \\ 1 & 0 & 1 \\ \ \end{matrix} \quad \begin{matrix} 0 & 1 & 0 \\ 1 & 0 & 0 \\ 0 & 0 & 1 \end{matrix} \right) \\
R_1(C_1(R_1(D_1, A_1, F_1))) &= \left( \begin{matrix} 0 & 1 \\ 1 & -3 \\ \ \end{matrix} \quad \begin{matrix} 1 & 0 & 1 \\ 0 & 0 & -3 \\ \ \end{matrix} \quad \begin{matrix} 0 & 1 & 0 \\ 1 & 0 & 0 \\ 0 & 0 & 1 \end{matrix} \right) \\
C_1(R_1(C_1(R_1(D_1, A_1, F_1)))) &= \left( \begin{matrix} 0 & 1 \\ 1 & -3 \\ \ \end{matrix} \quad \begin{matrix} 1 & 0 & 0 \\ 0 & 0 & -3 \\ \ \end{matrix} \quad \begin{matrix} 0 & 1 & 0 \\ 1 & 0 & -1 \\ 0 & 0 & 1 \end{matrix} \right) \\
&= (D_2, A_2, F_2) \text{ and} \\
C_2(R_2(C_2(R_2(D_2, A_2, F_2)))) &= \left( \begin{matrix} 0 & 1 \\ 1 & -3 \\ \ \end{matrix} \quad \begin{matrix} 1 & 0 & 0 \\ 0 & -3 & 0 \\ \ \end{matrix} \quad \begin{matrix} 0 & 0 & 1 \\ 1 & -1 & 0 \\ 0 & 1 & 0 \end{matrix} \right) \\
&= (D_3, A_3, F_3).
\end{aligned}$$

From the algorithm we conclude that $A$ is not regular.

Thus, we have given an algorithm to determine whether a given matrix over a Euclidean domain is regular or not and also a method of finding a $g$-inverse when it is regular.

## 4.5 Reflexive $g$-inverses of matrices

In the previous sections we looked at regular matrices and their $g$-inverses. Now we shall look at reflexive $g$-inverses.

We already know from section 3.1 that every regular matrix over a commutative ring has a reflexive $g$-inverse. We shall now find all reflexive $g$-inverses for a matrix over a principal ideal ring.

**Theorem 4.18:** Let $A$ be a regular matrix over a principal ideal ring.

(a) If $A = U\begin{bmatrix} I & 0 \\ 0 & 0 \end{bmatrix}V$ where $U$ and $V$ are unimodular matrices then every reflexive $g$-inverse of $A$ is of the form $V^{-1}\begin{bmatrix} I & B \\ C & CB \end{bmatrix}U^{-1}$ for some matrices $B$ and $C$ of appropriate order and every such matrix is a reflexive $g$-inverse of $A$.

(b) If $A = LR$ is a rank factorization of $A$ then $G = R_R^{-1}L_L^{-1}$ for some right inverse $R_R^{-1}$ or $R$ and some left inverse $L_L^{-1}$ of $L$, is a reflexive $g$-inverse of $A$ and also every reflexive $g$-inverse of $A$ is of this form for some $R_R^{-1}$ and $L_L^{-1}$.

(c) If $G_1$ and $G_2$ are $g$-inverses of $A$, $G_1AG_2$ is a reflexive $g$-inverse of $A$ and every reflexive $g$-inverse of $A$ is of the form $GAG$ for some $g$-inverse $G$ of $A$.

**Proof:** (a) Clearly every matrix $\begin{bmatrix} I & B \\ C & D \end{bmatrix}$ is a $g$-inverse of $\begin{bmatrix} I & 0 \\ 0 & 0 \end{bmatrix}$. If $\begin{bmatrix} I & B \\ C & D \end{bmatrix}$ is a reflexive $g$-inverse of $\begin{bmatrix} I & 0 \\ 0 & 0 \end{bmatrix}$ then,

$$\begin{bmatrix} I & B \\ C & D \end{bmatrix}\begin{bmatrix} I & 0 \\ 0 & 0 \end{bmatrix}\begin{bmatrix} I & B \\ C & D \end{bmatrix} = \begin{bmatrix} I & B \\ C & D \end{bmatrix},$$

i.e.,

$$\begin{bmatrix} I & B \\ C & CB \end{bmatrix} = \begin{bmatrix} I & B \\ C & D \end{bmatrix}.$$

This gives us $D = CB$. It can also be verified easily that $\begin{bmatrix} I & B \\ C & CB \end{bmatrix}$ is a reflexive $g$-inverse of $\begin{bmatrix} I & 0 \\ 0 & 0 \end{bmatrix}$. If $G$ is reflexive $g$-inverse of $U\begin{bmatrix} I & 0 \\ 0 & 0 \end{bmatrix}V$

then $VGU$ is a reflexive $g$-inverse of $\begin{bmatrix} I & 0 \\ 0 & 0 \end{bmatrix}$. Hence $VGU$ is of the form $\begin{bmatrix} I & B \\ C & CB \end{bmatrix}$ for some $B$ and $C$. Hence $G = V^{-1}\begin{bmatrix} I & B \\ C & CB \end{bmatrix}U^{-1}$. Conversely, it is easily verified that $V^{-1}\begin{bmatrix} I & B \\ C & CB \end{bmatrix}U^{-1}$ is a reflexive $g$-inverse of $U\begin{bmatrix} I & 0 \\ 0 & 0 \end{bmatrix}V$.

(b) Let $A = LR$ be a rank factorization. If $R_R^{-1}$ is a right inverse of $R$ and $L_L^{-1}$ is a left inverse of $L$ then $LRR_R^{-1}L_L^{-1}LR = LR$ and $R_R^{-1}L_L^{-1}LRR_R^{-1}L_L^{-1} = R_R^{-1}L_L^{-1}$ gives us that $R_R^{-1}L_L^{-1}$ is a reflexive $g$-inverse of $A$.

Conversely, let $G$ be a reflexive $g$-inverse of $A = LR$. Hence $LRGLR = LR$. By premultiplying with $L_L^{-1}$ and postmultiplying with $R_R^{-1}$ we get that $RGL = I$.

Hence $GL$ is an $R_R^{-1}$ and $RG$ is an $L_L^{-1}$. But $G = GAG = GLRG$. This gives us that $G = R_R^{-1}L_L^{-1}$ for some $R_R^{-1}$ and $L_L^{-1}$.

(c) is clear.

♠

The point of (c) is that a reflexive $g$-inverse of a regular matrix can be easily computed if one knows a $g$-inverse. In fact, if we know one $g$-inverse of a matrix $A$ then we can find all reflexive $g$-inverses of $A$ as the following result shows.

**Theorem 4.19:** Consider matrices over any commutative ring. If $G$ is a reflexive $g$-inverse of $A$ then every reflexive $g$-inverse of $A$ is of the form

$$G(X,Y) = G + (I - GA)XAG + GAY(I - AG) + (I - GA)XAY(I - AG)$$

for some $X, Y$ of appropriate order. Also every such $G(X,Y)$ is a reflexive $g$-inverse of $A$.

**Proof:** If $H$ is a reflexive $g$-inverse of $A$ then $H = G(H,H)$. On the other hand, by direct calculation one can verify that $G(X,Y)$ is a reflexive $g$-inverse of $A$ for every $X$ and $Y$.

♠

## 4.6 Some special integral domains

We have seen in section 4.3 that over a principal ideal domain every regular matrix is of the form $U\begin{bmatrix} I & 0 \\ 0 & 0 \end{bmatrix}V$ for some unimodular matrices $U$ and $V$ and that every regular matrix has a rank factorization.

In this section, we shall identify all integral domains which have the above properties. The results of this section will help us identify regular matrices over some natural rings and also help us in characterizing matrices which admit Moore-Penrose inverses.

**Theorem 4.20:** Over any integral domain the following are equivalent:

(i) every finitely generated projective module is free.

(ii) every regular matrix has a rank factorization.

(iii) every idempotent matrix has a rank factorization.

**Proof:** Let $D$ be the integral domain.

(i) $\Rightarrow$ (ii). Let every finitely generated projective module over $D$ be free. Let $A$ be an $m \times n$ regular matrix over $D$ of rank $r$. Consider $A$ as a module homomorphism from $D^n \to D^m$. Since $A$ is regular there exists a matrix $G : D^m \to D^n$ such that $AGA = A$. Clearly $AG$ is idempotent. Also $AG : D^m \to D^m$ is a module homomorphism with Range $(A) =$ Range $(AG) = \mathcal{S}$ (say). But for any idempotent linear map $T : D^m \to D^m$, range $(T)$ is projective (because, Range $(T) \oplus$ Range $(I - T) = D^m$). Hence we get that $\mathcal{S}$ is projective and by hypothesis it is free. Let $\mathcal{S}$ be isomorphic to $D^\ell$ for some integer $\ell$ through an isomorphism $\phi : \mathcal{S} \to D^\ell$. Let $C = \phi A$ and $B = i\phi^{-1}$ where $i : \mathcal{S} \to D^m$ is the inclusion map. Clearly $A = BC$. Also $B$ is an $m \times \ell$ matrix and $C$ is an $\ell \times n$ matrix.

Let us show that $A = BC$ is a rank factorization. Since $\mathcal{S}$ is isomorphic to $D^\ell, C : D^n \to D^\ell$ is surjective. Since $\mathcal{S}$ is a direct summand of $D^n$, there exists a matrix $C^0$ such that $CC^0$ is identity. Since $B$ is an injective map and $\mathcal{S}$ is a direct summand of $D^n$, there is a matrix $B^0$ such that $B^0B$ is identity. Now, since $CC^0 = I_\ell$ and $B^0B = I_\ell$, we get that $A = BC$ is a rank factorization.

Hence $A = BC$ is a rank factorization.

(ii) $\Rightarrow$ (iii) is clear because every idempotent matrix is regular.

(iii) $\Rightarrow$ (i). Let $X$ be a finitely generated projective module with $X \oplus Y = D^n$ for some module $Y$ and some integer $n$. Let $A : D^n \to D^n$ be the natural projection onto $X$ and let $\rho(A) = r$. $A$ is clearly idempotent. Let $A = CB$ be a rank factorization of $A$ where $C$ is an $n \times r$ matrix with a left inverse and $B$ is an $r \times n$ matrix with a right inverse. Hence $C : D^r \to D^n$ and range $(C) =$ range $(A) = X$. Since $C$ has a left inverse it follows that $X$ is free.

♠

Integral domains over which every finitely generated projective module is free behave as principal ideal rings, at least with respect to $g$-inverses, as the following Theorem demonstrates. We shall say that a commutative ring

$R$ is *projective free* if every finitely generated projective module over $R$ is free.

**Theorem 4.21:** For an integral domain $D$ the following are equivalent.

(i) $D$ is projective free.

(ii) Every idempotent matrix over $D$ is of the form $U\begin{bmatrix} I & 0 \\ 0 & 0 \end{bmatrix}U^{-1}$ for some unimodular matrix $U$.

(iii) Every regular matrix over $D$ is of the form $S\begin{bmatrix} I & 0 \\ 0 & 0 \end{bmatrix}T$ for some unimodular matrices $S$ and $T$.

(iv) Every regular matrix over $D$ has a rank factorization.

(v) Every idempotent matrix over $D$ has a rank factorization.

**Proof:** (i) $\Rightarrow$ (ii). By Theorem 4.20 we can assume that over the integral domain $D$ every regular matrix has a rank factorization. Now, let $A$ be an idempotent matrix over $D$. Since every idempotent matrix is regular, $A$ has a rank factorization, say $A = PQ$. If $A$ is of order $n \times n$ and rank $r$ then $P$ is of order $n \times r$ and $Q$ is of order $r \times n$. Now $I - A$ is also idempotent and has a rank factorization. Hence $\rho(I - A) = Tr(I - A) = n - r$. Thus, if $I - A = RS$ is a rank factorization then $R$ is of order $n \times n - r$ and $S$ is of order $n - r \times n$. Now the product $[P \;\; R]\begin{bmatrix} Q \\ S \end{bmatrix} = PQ + RS = A + I - A = I$. If we write $U = [P \;\; R]$ then $U$ is an $n \times n$ matrix and $U\begin{bmatrix} Q \\ S \end{bmatrix} = I$. Thus $U^{-1} = \begin{bmatrix} Q \\ S \end{bmatrix}$. Now,

$$U\begin{bmatrix} I & 0 \\ 0 & 0 \end{bmatrix}U^{-1} = [P \;\; R]\begin{bmatrix} I & 0 \\ 0 & 0 \end{bmatrix}\begin{bmatrix} Q \\ S \end{bmatrix} = PQ = A$$

this proves (ii).

(ii) $\Rightarrow$ (iii). Let $G$ be a $g$-inverse of a matrix $A$ of rank $r$. Hence $AG$ and $GA$ are idempotent. By (ii) we get unimodular matrices $U$ and $V$ such that $AG = U\begin{bmatrix} I & 0 \\ 0 & 0 \end{bmatrix}U^{-1}$ and $GA = V\begin{bmatrix} I & 0 \\ 0 & 0 \end{bmatrix}V^{-1}$. Hence

$$U^{-1}AV = U^{-1}AGAV = U^{-1}U\begin{bmatrix} I & 0 \\ 0 & 0 \end{bmatrix}U^{-1}AV = \begin{bmatrix} I & 0 \\ 0 & 0 \end{bmatrix}U^{-1}AV.$$

Also,

$$U^{-1}AV = U^{-1}AGAV = U^{-1}AV\begin{bmatrix} I & 0 \\ 0 & 0 \end{bmatrix}V^{-1}V = U^{-1}AV\begin{bmatrix} I & 0 \\ 0 & 0 \end{bmatrix}.$$

If we write $U^{-1}AV = \begin{bmatrix} B_1 & B_2 \\ B_3 & B_4 \end{bmatrix}$, then

$$\begin{bmatrix} B_1 & B_2 \\ B_3 & B_4 \end{bmatrix} = \begin{bmatrix} I & 0 \\ 0 & 0 \end{bmatrix} \begin{bmatrix} B_1 & B_2 \\ B_3 & B_4 \end{bmatrix} = \begin{bmatrix} B_1 & B_2 \\ 0 & 0 \end{bmatrix}.$$

Also,

$$\begin{bmatrix} B_1 & B_2 \\ B_3 & B_4 \end{bmatrix} = \begin{bmatrix} B_1 & B_2 \\ B_3 & B_4 \end{bmatrix} \begin{bmatrix} I & 0 \\ 0 & 0 \end{bmatrix} = \begin{bmatrix} B_1 & 0 \\ B_3 & 0 \end{bmatrix}.$$

Hence $U^{-1}AV = \begin{bmatrix} B_1 & 0 \\ 0 & 0 \end{bmatrix}$ and the order of $B_1$ equals order of $I$ which is $r \times r$. Hence $|B_1| \neq 0$. Since $A$ is regular, it follows that $B_1$ is regular. From the discussion after Theorem 3.6. it follows that $B_1$ is unimodular. Thus

$$A = U \begin{bmatrix} B_1 & 0 \\ 0 & 0 \end{bmatrix} V^{-1} = U \begin{bmatrix} I & 0 \\ 0 & 0 \end{bmatrix} \begin{bmatrix} B_1 & 0 \\ 0 & I \end{bmatrix} V^{-1}$$

where $\begin{bmatrix} B_1 & 0 \\ 0 & I \end{bmatrix}$ is or order $n \times n$. But $\begin{bmatrix} B_1 & 0 \\ 0 & I \end{bmatrix}$ is unimodular and hence $A$ is of the form $S \begin{bmatrix} I & 0 \\ 0 & 0 \end{bmatrix} T$ where $S$ and $T$ are unimodular matrices.

(iii) $\Rightarrow$ (i). We shall assume (iii) and prove (ii) of Theorem 4.20. If $A = S \begin{bmatrix} I & 0 \\ 0 & 0 \end{bmatrix} T$, let $A = [S_1 \; S_2] \begin{bmatrix} I & 0 \\ 0 & 0 \end{bmatrix} \begin{bmatrix} T_1 \\ T_2 \end{bmatrix}$.

Then, $A = S_1 T_1$. Since $S$ and $T$ are unimodular it follows that $S_1$ has a left inverse and $T_1$ has a right inverse.

The equivalence of (iv) and (v) with (i) was already proved in the previous Theorem.

♠

From the above Theorem it follows that if $A$ is an $m \times n$ matrix with a right inverse over a projective free integral domain, then there is a matrix $B$ such that $\begin{bmatrix} A \\ B \end{bmatrix}$ is unimodular. Indeed, this follows from (iii) of the previous Theorem. Let us remark that we shall generalize the previous Theorem to arbitrary commutative rings in Theorem 8.15.

We shall also characterize symmetric idempotent matrices which will be needed in the next section.

**Proposition 4.22:** Let $R$ be an integral domain over which every finitely generated projective module is free. Let $A$ be a symmetric idempotent matrix of rank $r$ over $R$. Then there exists a unimodular matrix $U$

such that

$$A = U \begin{bmatrix} I & 0 \\ 0 & 0 \end{bmatrix} U^{-1}$$

where $U^T U = \begin{bmatrix} B & 0 \\ 0 & C \end{bmatrix}$ for some matrices $B$ and $C$.

**Proof:** Since $A$ is idempotent, by the previous Theorem, there is a unimodular matrix $U$ such that

$$A = U \begin{bmatrix} I & 0 \\ 0 & 0 \end{bmatrix} U^{-1}.$$

Since $A$ is symmetric, $U \begin{bmatrix} I & 0 \\ 0 & 0 \end{bmatrix} U^{-1} = (U^T)^{-1} \begin{bmatrix} I & 0 \\ 0 & 0 \end{bmatrix} U^T$.

Hence $U^T U \begin{bmatrix} I & 0 \\ 0 & 0 \end{bmatrix} = \begin{bmatrix} I & 0 \\ 0 & 0 \end{bmatrix} U^T U$.

If $U^T U = \begin{bmatrix} B & D \\ E & C \end{bmatrix}$ then it follows that $D = 0 = E$. Hence the result.

♠

There are many natural projective free rings, thanks to Serre's conjecture and Quillen-Souslin Theorem – for any principal ideal ring $R$, the ring of polynomials $R[x_1, x_2, \cdots, x_n]$ is one such (see [55] and [51]). We shall use this fact in the next section.

## 4.7 Examples

We shall look at matrices over many natural rings. For many of these rings we shall assume that the identity is the involution while dealing with Moore-Penrose inverses. But the results can be modified suitably if the involution is not the identity involution. Over these rings we shall also characterize all those matrices which have Moore-Penrose inverses. Recall that $G$ is called a Moore-Penrose inverse of $A$ if $A$ and $G$ satisfy the Moore-Penrose equations (1)–(4). If $A$ and $G$ satisfy (1) and (3) then $AG$ is a symmetric idempotent matrix. For the natural rings which we are going to look at we shall find ways of characterizing symmetric idempotent matrices.

Sometimes we shall also explicitly characterize matrices which have $\{1,3\}$-inverses and $\{1,4\}$-inverses. We shall start with the simplest possible principal ideal ring which is not a field.

**Example 1:** $\mathbb{Z}$, the ring of integers. Over this ring we already know the structure of regular matrices from section 4.3.

We shall now look at Moore-Penrose inverses.

Following [86], We say that an associative ring $R$ with an involution $a \to \bar{a}$ satisfies the *Rao condition* if $a_1 = a_1\bar{a}_1 + a_2\bar{a}_2 + \cdots + a_n\bar{a}_n$ only if $a_2 = a_3 = \cdots = a_n = 0$

We will be mainly interested in integral domains satisfying the Rao condition. The ring of integers, $\mathbb{Z}$ with the identity involution $a \to \bar{a} = a$ satisfies the Rao condition. On the other hand, a field with the identity involution never satisfies the Rao condition. Indeed, in any field $1 = 1^2 + 1^2 + 1^2$ if characteristic$(F) = 2$ and $\frac{1}{2} = (\frac{1}{2})^2 + (\frac{1}{2})^2$ if characteristic $F \neq 2$. Over the integral domains that satisfy the Rao condition we can characterize symmetric idempotent matrices.

**Proposition 4.23:** Let $A$ be a symmetric idempotent matrix over an integral domain $R$ satisfying the Rao condition. Then there exists a permutation matrix $P$ such that $A = P\begin{bmatrix} I & 0 \\ 0 & 0 \end{bmatrix}P^T$.

**Proof:** Since $A$ is symmetric and idempotent $AA^T = A^2 = A$. If $(a_1\ a_2\ \cdots\ a_n)$ is the first row of $A$ then $a_1^2 + \cdots + a_n^2 = a_1$. By the Rao condition we get that $a_2 = \cdots = a_n = 0$ and $a_1^2 = a_1$. Since $R$ is an integral domain we get that $a_1 = 0$ or 1. Considering all the rows of $A$ gives us that $A$ is a diagonal matrix and that diagonal elements are 0 or 1. It is clear that there is a permutation matrix $P$ such that $A = P\begin{bmatrix} I & 0 \\ 0 & 0 \end{bmatrix}P^T$.

♠

We can now characterize all matrices $A$ that admit Moore-Penrose inverses over an integral domain which satisfies the Rao condition.

**Theorem 4.24:** Let $A$ be a matrix over an integral domain satisfying the Rao condition. Then

(a) $A^+$ exists if and only if there are permutation matrices $P$ and $Q$ and a unimodular matrix $M(A)$ such that $A = P\begin{bmatrix} M(A) & 0 \\ 0 & 0 \end{bmatrix}Q^T$. In this case $Q\begin{bmatrix} M(A)^{-1} & 0 \\ 0 & 0 \end{bmatrix}P^T$ is $A^+$.

(b) $A_{13}^-$ exists if and only if there is a permutation matrix $P$ such that $A = P\begin{bmatrix} B \\ 0 \end{bmatrix}$ where $B$ has a right inverse. If $B_R^{-1}$ is a right inverse of $B$ then $[B_R^{-1}\ 0]P^T$ is a $\{1,3\}$-inverse of $A$.

(c) $A_{14}^-$ exists if and only if there is a permutation matrix $Q$ such that $A = [B\ \ 0]Q$ where $B$ has a left inverse. If $B_L^{-1}$ is a left inverse of $B$ then $Q^T\begin{bmatrix} B_L^{-1} \\ 0 \end{bmatrix}$ is a $\{1,4\}$-inverse of $A$.

**Proof:** (a) Let $A^+$ exist. Then $AA^+$ and $A^+A$ are both symmetric idempotent matrices. Hence, by Proposition 4.23 there are permutation matrices $P$ and $Q$ such that

$$AA^+ = P\begin{bmatrix} I & 0 \\ 0 & 0 \end{bmatrix}P^T \text{ and } A^+A = Q\begin{bmatrix} I & 0 \\ 0 & 0 \end{bmatrix}Q^T.$$

Since $\rho(AA^+) = \rho(A^+A)$ $(= \rho(A) = \rho(A^+))$ over any commutative ring, $I$ in the above matrices are both of order $r \times r$ where $r = \rho(A)$.

Now, as in the proof of (ii) $\Rightarrow$ (iii) of Theorem 4.21

$$\begin{aligned} P^TAQ = P^TAA^+AQ &= P^TAA^+PP^TAQ \\ &= \begin{bmatrix} I & 0 \\ 0 & 0 \end{bmatrix}P^TAQ \end{aligned}$$

Also,

$$\begin{aligned} P^TAQ = P^TAA^+AQ &= P^TAQQ^TA^+AQ \\ &= P^TAQ\begin{bmatrix} I & 0 \\ 0 & 0 \end{bmatrix}. \end{aligned}$$

If we write $P^TAQ$ as the partitioned matrix $\begin{bmatrix} B & C \\ D & E \end{bmatrix}$ where $B$ is an $r \times r$ matrix then substituting this in the above equations gives us that $C = D = E = 0$. Thus $P^TAQ = \begin{bmatrix} B & 0 \\ 0 & 0 \end{bmatrix}$.

In the same way as above, since the relative roles of $A$ and $A^+$ are symmetric, we get that $Q^TA^+P = \begin{bmatrix} F & 0 \\ 0 & 0 \end{bmatrix}$ for some $r \times r$ matrix $F$.

Since $P^TAA^+P = P^TAQQ^TA^+P = \begin{bmatrix} B & 0 \\ 0 & 0 \end{bmatrix}\begin{bmatrix} F & 0 \\ 0 & 0 \end{bmatrix} = \begin{bmatrix} I & 0 \\ 0 & 0 \end{bmatrix}$ we get that $BF = I$ and $B$ is a unimodular matrix. If we call $B$ as $M(A)$ we get that $A = P\begin{bmatrix} M(A) & 0 \\ 0 & 0 \end{bmatrix}Q^T$ where $M(A)$ is a unimodular matrix and $A^+ = Q\begin{bmatrix} M(A)^{-1} & 0 \\ 0 & 0 \end{bmatrix}P^T$.

(b) Let $A$ be such that $A^-_{13}$ exists. By looking at $A^-_{13}AA^-_{13}$ we see that, in fact, $A^-_{13}AA^-_{13}$ is a $\{1,2,3\}$-inverse. Let $G = A^-_{13}AA^-_{13}$. Then $AG$ is symmetric and idempotent.

Hence by Proposition 4.23 it follows that there is a permutation matrix $P$ such that $AG = P\begin{bmatrix} I & 0 \\ 0 & 0 \end{bmatrix}P^T$. If $r = \rho(A)$ since $\rho(AG) = \rho(A)$ we get that $I$ is an $r \times r$ matrix.

Now, again as in the proof of (ii) $\Rightarrow$ (iii) of Theorem 4.21,

$$P^TA = P^TAGPP^TA = \begin{bmatrix} I & 0 \\ 0 & 0 \end{bmatrix} P^TA.$$

Also, $GP = GAGP = GPP^TAGP = GP\begin{bmatrix} I & 0 \\ 0 & 0 \end{bmatrix}$. If we write $P^TA = \begin{bmatrix} B_1 \\ B_2 \end{bmatrix}$ then $\begin{bmatrix} B_1 \\ B_2 \end{bmatrix} = \begin{bmatrix} I & 0 \\ 0 & 0 \end{bmatrix}\begin{bmatrix} B_1 \\ B_2 \end{bmatrix} = \begin{bmatrix} B_1 \\ 0 \end{bmatrix}$.

Thus $P^TA$ is of the form $\begin{bmatrix} B \\ 0 \end{bmatrix}$. Similarly $GP$ is of the form $[F \;\; 0]$ and $P^TAGP = \begin{bmatrix} B \\ 0 \end{bmatrix}[F \;\; 0] = \begin{bmatrix} BF & 0 \\ 0 & 0 \end{bmatrix} = \begin{bmatrix} I & 0 \\ 0 & 0 \end{bmatrix}$.

Hence $A = P\begin{bmatrix} B \\ 0 \end{bmatrix}$ where $B$ has a right inverse and every $\{1,2,3\}$-inverse is of the form $[F \;\; 0]P^T$ where $F$ is a right inverse of $B$.

(c) can be proved similarly.

♠

The previous Theorem characterizes all those matrices over an integral domain satisfying the Rao condition that have Moore-Penrose inverse. The characterization is that the submatrix of $A$ formed by removing the zero rows and zero columns must have an inverse. Similarly, for the existence of $\{1,3\}$-inverse, the submatrix of $A$ formed by removing the zero rows of $A$ must have a right inverse.

Thus we have a complete characterization of integer matrices which have Moore-Penrose inverse, $\{1,3\}$-inverses and $\{1,4\}$-inverses. Let us look at another example.

**Example 2:** $\mathbb{Z}[x_1, x_2, \cdots, x_n]$, the ring of polynomials in the variables $x_1, x_2, \cdots, x_n$ with integer coefficients.

$\mathbb{Z}[x_1, x_2, \cdots, x_n]$ is an integral domain and satisfies the condition that every finitely generated projective module over this ring is free (see [55] and [51]). $\mathbb{Z}[x_1, x_2, \cdots, x_n]$ is not a principal ideal domain.

Thus, regular matrices over $\mathbb{Z}[x_1, x_2, \cdots, x_n]$ are characterized by Theorem 4.21. Matrices over $\mathbb{Z}[x_1, x_2, \cdots, x_n]$ that admit Moore-Penrose inverse, $\{1,3\}$-inverse and $\{1,4\}$-inverse are characterized by Theorem 4.24 because $\mathbb{Z}[x_1, x_2, \cdots, x_n]$ satisfies the Rao condition. This follows from the following Proposition.

**Proposition 4.25:** For an integral domain $R$ with an involution $a \to \bar{a}$, if $R$ satisfies the Rao condition then the ring of polynomials $R[x]$ also satisfies the Rao condition.

**Proof:** Observe that if $R$ satisfies the Rao condition then whenever $0 = a_1\bar{a}_1 + a_2\bar{a}_2 + \cdots + a_n\bar{a}_n$ all the $a_1, a_2, \cdots, a_n$ are $= 0$.

Now let $a_1 = a_1\bar{a}_1 + a_2\bar{a}_2 + \cdots + a_n\bar{a}_n$ in $R[x]$. If all $a_1, a_2, \cdots, a_n$ are of degree $= 0$ then there is nothing to verify. If at least one of the degrees of $a_1, a_2, \cdots, a_n$ is $>0$ then we get an equation of the type $0 = b_1\bar{b}_1 + b_2\bar{b}_2 + \cdots + b_n\bar{b}_n$ by looking at the highest (say, $k$) among the degrees of $a_1, a_2, \cdots, a_n$ and equating the coefficients of $x^k$ in $a_1 = a_1\bar{a}_1 + a_2\bar{a}_2 + \cdots + a_n\bar{a}_n$ in $R[x]$. This gives us a contradiction.

♠

**Example 3:** $I\!R[x]$, the ring of polynomials in the variable $x$ with real coefficients.

$I\!R[x]$ is a principal ideal ring and hence the regular matrices are characterized by Theorem 4.13. Let us now characterize matrices over $I\!R[x]$ that admit Moore-Penrose inverse.

An important property satisfied by $I\!R$ is

$$(*) \qquad \sum_1^n a_i^2 = 0 \Rightarrow a_i = 0 \ \text{ for all } \ i.$$

We shall start with a characterization of symmetric idempotent matrices.

**Proposition 4.26:** Every symmetric idempotent matrix over $I\!R[x]$ is a real matrix. If $B$ is a symmetric idempotent matrix over $I\!R[x]$ then there exists an orthogonal matrix $U$ over $I\!R$ such that $B = U \begin{bmatrix} I & 0 \\ 0 & 0 \end{bmatrix} U^T$.

**Proof:** Let $B$ be a symmetric idempotent matrix over $I\!R[x]$. Let $B = B_0 + B_1x + \cdots + B_nx^n$ where $B_0, B_1, \cdots, B_n$ are matrices over $I\!R$ and $B_n \neq 0$. Let if possible, $n \geq 1$. Since $B$ is symmetric and idempotent we have that $B_n^TB_n = B_n^2 = 0$. But by $(*)$ we have that $B_n = 0$, a contradiction. Hence $B$ is real. The rest is clear.

♠

We are now ready to characterize matrices over $I\!R[x]$ that admit Moore-Penrose and other inverses.

**Theorem 4.27:** Let $A$ be a matrix of rank $r$ over $I\!R[x]$. Then

(a) $A^+$ exists if and only if there exist real orthogonal matrices $U$ and $V$ and an $r \times r$ unimodular matrix $M$ over $I\!R[x]$ such that

$A = U \begin{bmatrix} M & 0 \\ 0 & 0 \end{bmatrix} V$. In this case $V^T \begin{bmatrix} M^{-1} & 0 \\ 0 & 0 \end{bmatrix} U^T$ is $A^+$.

(b) $A$ has a $\{1,3\}$-inverse if and only if there exists a real orthogonal matrix $U$ such that $A = U\begin{bmatrix} B \\ 0 \end{bmatrix}$ where $B$ has a right inverse over $\mathbb{R}[x]$. If $B_R^{-1}$ is a right inverse of $B$ then $[B_R^{-1}\ 0]U^T$ is a $\{1,3\}$-inverse.

(c) $A$ has a $\{1,4\}$-inverse if and only if there exists a real orthogonal matrix $V$ such that $A = [B\quad 0]V$ where $B$ has a left inverse over $\mathbb{R}[x]$. If $B_L^{-1}$ is a left inverse of $B$ then $V^T\begin{bmatrix} B_L^{-1} \\ 0 \end{bmatrix}$ is a $\{1,4\}$-inverse.

**Proof:** We shall prove only (a).

Let $A^+$ exist. Then $AA^+$ and $A^+A$ are both symmetric idempotent matrices. Hence by Proposition 4.26 there exist orthogonal matrices $U$ and $V$ such that

$$AA^+ = U\begin{bmatrix} I & 0 \\ 0 & 0 \end{bmatrix}U^T \text{ and } A^+A = V\begin{bmatrix} I & 0 \\ 0 & 0 \end{bmatrix}V^T.$$

Following exactly as in the proof of (a) of Theorem 4.24 we get that $A = U\begin{bmatrix} M & 0 \\ 0 & 0 \end{bmatrix}V^T$ for some $r \times r$ unimodular matrix $M$ over $\mathbb{R}[x]$ and $A^+ = V\begin{bmatrix} M^{-1} & 0 \\ 0 & 0 \end{bmatrix}U^T$.

♠

The $M$ above is unique upto multiplication by orthogonal matrices.

Since $\mathbb{R}[x]$ is a very useful ring when it comes to applications we now comment on some computational aspects involving matrices over this ring.

The algorithm given for Euclidean domains is applicable over $\mathbb{R}[x]$ and this gives an algorithm for verifying whether a given matrix is regular or not and find a $g$-inverse if it is regular. To give an algorithm to verify as to when a matrix $A$ over $\mathbb{R}[x]$ admits the Moore-Penrose inverse we have,

**Theorem 4.28:** For a matrix $A$ over $\mathbb{R}[x], A^+$ exists if and only if any of $AA^TAA^T, A^TAA^TA$ and $A^TAA^T$ admit $g$-inverses.

In this case $A^+$ equals any of

$A^T(AA^TAA^T)^-AA^T, A^TA(A^TAA^TA)^-A^T$ and $A^T(A^TAA^T)^-A^T$.

**Proof:** $\mathbb{R}[x]$ satisfies $(*)$. Hence $BB^TC = BB^TD$ implies that $B^TC = B^TD$. This, as in the classical case (See [83], page 57) gives the "if" part. The only if part follows from Theorem 4.27.

♠

The above Theorem gives the required algorithm for $A^+$. The corresponding results for $A_{13}^-$ and $A_{14}^-$ can be formulated and proved easily.

**Example 4:** The ring $\mathbb{R}[x_1, x_2, \cdots, x_n]$ of polynomials in variables $x_1, x_2, \cdots, x_n$ with coefficients from $\mathbb{R}$.

Though $\mathbb{R}[x_1, x_2, \cdots, x_n]$ is not a principal ideal domain, since every finitely generated projective module over $\mathbb{R}[x_1, x_2, \cdots, x_n]$ is free by Quillen-Souslin Theorem the results of Theorem 4.21 are applicable. Hence all regular matrices are characterized. $\mathbb{R}[x_1, x_2, \cdots, x_n]$ also satisfies $(*)$ of the previous example. Hence the arguments of the previous example go through and we have,

**Theorem 4.29:** For a matrix $A$ of rank $r$ over $\mathbb{R}[x_1, x_2, \cdots, x_n]$, $A^+$ exists if and only if there exists real orthogonal matrices $U$ and $V$ and a unimodular $r \times r$ matrix $M$ such that

$$A = U \begin{bmatrix} M & 0 \\ 0 & 0 \end{bmatrix} V$$

In this case $A^+ = V^T \begin{bmatrix} M^{-1} & 0 \\ 0 & 0 \end{bmatrix} U^T$.

♠

$\{1,3\}$-inverses and $\{1,4\}$-inverses can be dealt with in a similar way. Regarding various computations it is possible to give algorithms using Gröbner bases ([97]).

**Example 5:** The ring $F[x_1, x_2, \cdots, x_n]$ of polynomials in variables $x_1, x_2, \cdots, x_n$ with coefficients from a field $F$.

Though $F[x_1, x_2, \cdots, x_n]$ is not a principal ideal ring Theorem 4.21 is still applicable. Hence all regular matrices are characterized.

Let us now look at Moore-Penrose inverses.

**Theorem 4.30:** Let $A$ be a matrix of rank $r$ over $F[x_1, x_2, \cdots, x_n]$. Then

(a) $A^+$ exists if and only if there exists unimodular matrices $U$ and $V$ and an $r \times r$ unimodular matrix $M$ such that

$$A = U \begin{bmatrix} M & 0 \\ 0 & 0 \end{bmatrix} V, \quad U^T U = \begin{bmatrix} B & 0 \\ 0 & C \end{bmatrix} \text{ and } VV^T = \begin{bmatrix} D & 0 \\ 0 & E \end{bmatrix}$$

for some $B, C, D$ and $E$.

In this case $A^+ = V^{-1} \begin{bmatrix} M^{-1} & 0 \\ 0 & 0 \end{bmatrix} U^{-1}$.

(b) $A$ has a $\{1,3\}$-inverse if and only if there exists a unimodular matrix $U$ such that $A = U \begin{bmatrix} F \\ 0 \end{bmatrix}$ where $F$ has a right inverse and $U^T U = \begin{bmatrix} B & 0 \\ 0 & C \end{bmatrix}$ for some $B$ and $C$.

(c) $A$ has a $\{1,4\}$-inverse if and only if there exists a unimodular matrix $V$ such that $A = [F\ \ 0]V$ where $F$ has a left inverse and $VV^T = \begin{bmatrix} B & 0 \\ 0 & C \end{bmatrix}$ for some $B$ and $C$.

**Proof:** The proof is similar to the proofs of Theorems 4.24 and 4.27. One has to use Proposition 4.22.

♠

# EXERCISES

**Exercise 4.1:** Over an associative ring show that a matrix $A = \begin{bmatrix} S & 0 \\ 0 & 0 \end{bmatrix}$ where $S = diag\ (s_1, s_2, \cdots, s_r)$ with $s_i \neq 0$ for $1 \leq i \leq r$ is regular if and only if $s_1, s_2, \cdots, s_r$ are all regular in $R$. Also show that a matrix $G$ is a $g$-inverse of $A$ if and only if it is of the form $\begin{bmatrix} S^- & B \\ C & D \end{bmatrix}$ where $S^-$ is a $g$-inverse of $S$. Show however that not every $g$-inverse of a diagonal matrix all of whose diagonal elements are non zero need be a diagonal matrix (Hint: consider $A = \begin{bmatrix} 2 & 0 \\ 0 & 4 \end{bmatrix}$ over $\mathbb{Z}[6]$.)

**Exercise 4.2:** Over an associative ring show that a matrix $A = \begin{bmatrix} T & 0 \\ 0 & 0 \end{bmatrix}$ where $T$ is an $r \times r$ matrix is regular if and only if $T$ is regular. Also show that a matrix $G$ is a $g$-inverse of $A$ if and only if it is of the form $\begin{bmatrix} T^- & B \\ C & D \end{bmatrix}$ where $T^-$ is a $g$-inverse of $T$.

**Exercise 4.3:** Over an integral domain show that a $g$-inverse $G$ of $A$ is a reflexive $g$-inverse of $A$ if and only if $\rho(G) = \rho(A)$. Show also that this result is not true for matrices over general commutative rings.

**Exercise 4.4:** Give an example of a commutative ring with 1 such that not every matrix has a rank factorization.

**Exercise 4.5:** With reference to (b) of Theorem 4.3, let (b′) be: If $d_1, d_2, \cdots$ is a sequence of elements from $R$ such that $d_{i+1}|d_i$ for all $i \geq 1$ then there is an $n$ such that $d_n|d_{n+1}$. Over a commutative ring with 1 are (b) and (b′) equivalent? Note that it is the property (b′) of principal ideal rings which is required in the proof of the Smith Normal Form Theorem in section 4.2.

**Exercise 4.6:** Call a unimodular matrix an essentially $2 \times 2$ unimodular matrix if it looks like $\begin{bmatrix} V & 0 \\ 0 & I \end{bmatrix}$ after permuting the rows and columns where $V$ is a $2 \times 2$ unimodular matrix. Show that every unimodular matrix is a product of essentially $2 \times 2$ unimodular matrices. This gives a representation of all unimodular matrices.

**Exercise 4.7:** Examine the veracity of Proposition 4.25 for general commutative rings.

**Exercise 4.8:** The Rao condition looks a little biased towards $a_1$. Let us set things right in the following.

Let $R$ be an associative ring with 1 and with an involution $a \to \bar{a}$. Show that the following are equivalent.

(i) $R$ satisfies the Rao condition.

(ii) If $1 = \Sigma a_i \bar{a}_i$ in $R$ then $a_i = 0$ for every $i$ except possibly for one $i$.

(iii) If $E$ is an $n \times n$ matrix such that $E^2 = E = E^*$ then $E$ is *diag* $(e_1, e_2, \cdots, e_n)$ where $e_i^2 = e_i = \bar{e}_i$ for all $i$.

(iv) If $u = \Sigma a_i \bar{a}_i$ is a unit in $R$ then $\bar{a}_i u^{-1} a_j = 0$ whenever $i \neq j$.

**Exercise 4.9:** Let $R$ be an associative Ring with 1 and an involution $a \to \bar{a}$. Then show that the following are equivalent.

(i) $R$ satisfies the Rao condition and 0 and 1 are the only elements $a$ such that $a\bar{a} = a$.

(ii) An $m \times n$ matrix $A$ over $R$ has Moore-Penrose inverse if and only if the nonzero part of $A$ (i.e., the submatrix of $A$ obtained by removing the zero rows and zero columns of $A$) is a square matrix and is invertible.

**Exercise 4.10:** (a) Show that the following rings satisfy the Rao condition.

(1) The ring of Gaussian integers $\mathbb{Z}[i]$ with $m + ni \to m - ni$ as the involution.

(2) The ring $\mathbb{Z}[(1+\sqrt{-3})/2]$ with $a+b\sqrt{-3} \to a-b\sqrt{-3}$ as the involution.

(b) Show that the following rings do not satisfy the Rao condition.

(1) the field of complex numbers $\mathbb{C}$ with $a+bi \to a-bi$ as the involution.

(2) the ring of Gaussian integers $\mathbb{Z}[i]$ with $m + ni \to m + ni$ as the involution.

(3) The ring of all $2 \times 2$ matrices over a field $F$ with the involution $\begin{bmatrix} a & b \\ c & d \end{bmatrix} \to \begin{bmatrix} d & -b \\ -c & a \end{bmatrix}$.

**Exercise 4.11:** Study Moore-Penrose inverses of matrices over $F[x]$ for a formally real field $F$. Recall that a Field is said to be *formally real* if $\Sigma_1^k \alpha_i^2 = 0, \alpha_i \in F$ implies that $\alpha_i = 0$ for all $i$.

The exact statement of Theorem 4.27 is not true for $F[x]$ when $F$ is only a formally real field. However the correct analogous result is: $A^+$ exists if and only if $A = P_1 M Q_1$ where $P_1$ is an $m \times r$ matrix over $F$ such that $P_1^T P_1 = \Delta$ is a diagonal matrix of rank $r$, $Q_1$ is an $r \times n$ matrix over $F$ such that $Q_1 Q_1^T = \Gamma$ is a diagonal matrix of rank $r$ and $M$ is an $r \times r$ unimodular matrix over $F[x]$. In this case $Q_1^T \Gamma^{-1} M^{-1} \Delta^{-1} P_1^T$ is $A^+$.

**Exercise 4.12:** Study Moore-Penrose inverses of matrices over $F[x]$ for a real closed field $F$. Recall that a *real closed field* is a formally real field in which every element is either a square or the negative of a square and every polynomial over $F$ in a variable $x$ of odd degree has a root in $F$. The exact statement of Theorem 4.27 is true for matrices over $F[x]$ where $F$ is a real closed field.

**Exercise 4.13:** Show that for matrices over $F[x]$ where $F$ is formally real the concepts of minimum-norm $g$-inverse, the least squares $g$-inverse and minimum norm least squares $g$-inverse make sense and that they coincide with $\{1,4\}$-inverse, $\{1,3\}$-inverse and $A^+$ respectively (see [83] for the case of real matrices).

Observe that any formally real field $F$ is ordered and this order extends to $F[x]$ also and makes it an ordered ring. We can also define an $F$-valued norm for vectors in $(F[x])^n$. The proof of the above result can be obtained as in the real case by following Theorems 3.1.1, 3.2.1 and 3.3.1 of Rao and Mitra [83].

**Exercise 4.14:** Study Moore-Penrose inverses of matrices over $F[x]$ for a field $F$ of characteristic $\neq 2$. Use Theorem 4.7 of [64].

**Exercise 4.15:** Characterize integer matrices having group inverse. Characterize also integer matrices having Drazin inverse.

**Exercise 4.16:** In any associative ring $R$ with 1, for elements $a$ and $b$ say that $a$ and $b$ are equivalent if there are units $p$ and $q$ such that $paq = b$ and say that $a$ and $b$ are similar if there is a unit $u$ such that $uau^{-1} = b$. Prove the following results (see Song and Guo [92]).

(a) $a$ and $b$ are idempotent elements in $R$ show that $a$ and $b$ are equivalent if and only if they are similar.

(b) If $A$ is an idempotent matrix over $R$ and is equivalent to a diagonal matrix then show that it is similar to a diagonal matrix. Compare with Theorem 4.21.

# Chapter 5

# Regularity – basics

We now come to a preliminary but basic study of regular matrices over commutative rings. These results will be used in the next Chapter. While dealing with matrices over commutative rings also we shall use these results.

## 5.1 Regularity of rank one matrices

In Theorem 2.5 we have seen that for a matrix $A$ of rank $r$ over a commutative ring the compound matrix $C_r(A)$ is of rank 1. Also, from Theorem 2.4 (b) it follows that $C_r(G)$ is a $g$-inverse of $C_r(A)$ whenever $G$ is a $g$-inverse of $A$. In view of these two results we shall first decide conditions for the regularity of a rank one matrix. We shall also find all $g$-inverses of a rank one regular matrix.

**Theorem 5.1:** Let $A = (a_{ij})$ be an $m \times n$ matrix over a commutative ring with $\rho(A) = 1$. Then,

(a) $A$ is regular if and only if there exists elements, $\{g_{ji} : 1 \leq j \leq n,\ 1 \leq i \leq m\}$ in $R$ such that $\left(\sum\limits_{i,j} a_{ij}g_{ji}\right) a_{k\ell} = a_{k\ell}$.

(b) If $\{g_{ij} : 1 \leq i \leq n,\ 1 \leq j \leq m\}$ are elements in $R$ such that $\left(\sum\limits_{i,j} a_{ij}g_{ji}\right) a_{k\ell} = a_{k\ell}$ then the matrix $G = (g_{ij})$ is a $g$-inverse of $A$.

Also every $g$-inverse $G = (g_{ij})$ of $A$ satisfies $\left(\sum\limits_{i,j} a_{ij}g_{ji}\right) a_{k\ell} = a_{k\ell}$.

**Proof:** We shall prove (a) and (b) simultaneously. Let $G = (g_{ij})$ be a $g$-inverse of $A$, i.e., $AGA = A$. This implies that $\sum\limits_{j,i} a_{kj}g_{ji}a_{i\ell} = a_{k\ell}$ for all $k$

and $\ell$. Since $\rho(A) = 1$, we have that every $2 \times 2$ subdeterminant of $A$ is zero. Hence

$$\begin{vmatrix} a_{kj} & a_{k\ell} \\ a_{ij} & a_{i\ell} \end{vmatrix} = 0$$

for every $k, j, i$ and $\ell$ (even with repetition of some of the suffixes). Hence $a_{kj}a_{i\ell} = a_{ij}a_{k\ell}$ for all $k, j, i$ and $\ell$. Thus $\left(\sum_{i,j} a_{ij}g_{ji}\right) a_{k\ell} = a_{k\ell}$ for all $k$ and $\ell$.

Conversely, if $\left(\sum_{i,j} a_{ij}g_{ji}\right) a_{k\ell} = a_{k\ell}$ for all $k$ and $\ell$, since $\rho(A) = 1$ we get that $\sum_{j,i} a_{kj}g_{ji}a_{i\ell} = a_{k\ell}$ for all $k$ and $\ell$. Thus $AGA = A$, i.e., $G = (g_{ji})$ is a $g$-inverse of $A$.

♠

It is only a small step to formulate the result of the previous Theorem for matrices over integral domains.

**Theorem 5.2:** Let $A = (a_{ij})$ be an $m \times n$ matrix and $G = (g_{ji})$ be an $n \times m$ matrix over an integral domain. Let further $\rho(A) = 1$. Then the following are equivalent

(i) $AGA = A$.

(ii) $\sum_{i,j} a_{ij}g_{ji} = 1$, i.e., $Tr(AG) = 1$.

♠

Theorem 5.1 leads us to the most basic result of this monograph. We present this in the next section.

## 5.2 A basic result on regularity

Continuing with the arguments of the first paragraph of the previous section, in the following Theorem, we shall give some necessary and some sufficient conditions for a general matrix over a commutative ring to be regular. This turns out to be the most basic result of this monograph.

Let us recall from section 2.1 that $Q_{r,m}$ stands for the set $\{\alpha : \alpha = \{\alpha_1, \alpha_2, \cdots, \alpha_r\}, 1 \leq \alpha_1 < \alpha_2 < \cdots < \alpha_r \leq m\}$ and $\Delta_{r,m,n}$ stands for the set $Q_{r,m} \times Q_{r,n}$. We shall write $\Delta$ for $\Delta_{r,m,n}$ if there is no scope for confusion. Our basic Theorem is the following:

**Theorem 5.3:** Let $A = (a_{ij})$ be an $m \times n$ matrix over a commutative ring $R$ with $\rho(A) = r$. Consider the statements:

(i) There exist elements $\{c_{\beta\alpha} : (\alpha,\beta) \in \Delta\}$ in $R$ such that $\sum_{(\alpha,\beta)\in\Delta} |A_{\alpha\beta}|c_{\beta\alpha} = 1$.

(ii) There exist elements $\{c_{\beta\alpha} : (\alpha,\beta) \in \Delta\}$ in $R$ such that $(\sum_{(\alpha,\beta)\in\Delta} |A_{\alpha\beta}|c_{\beta\alpha})a_{k\ell} = a_{k\ell}$ for all $k$ and $\ell$.

(iii) $A$ is regular.

(iv) $C_r(A)$ is regular.

(v) There exist elements $\{c_{\beta\alpha} : (\alpha,\beta) \in \Delta\}$ in $R$ such that $|A_{\gamma\delta}|(\sum_{(\alpha,\beta)\in\Delta} |A_{\alpha\beta}|c_{\beta\alpha}) = |A_{\gamma\delta}|$ for all $(\gamma,\delta) \in \Delta$.

(vi) There exist elements $\{c_{\beta\alpha} : (\alpha,\beta) \in \Delta\}$ in $R$ such that $\sum_{(\alpha,\beta)\in\Delta} |A_{\alpha\beta}|c_{\beta\alpha}$ is a nonzero idempotent.

Then, (i) $\Rightarrow$ (ii) $\Rightarrow$ (iii) $\Rightarrow$ (iv) $\Rightarrow$ (v) $\Rightarrow$ (vi).

In case 0 and 1 are the only idempotents in $R$, (vi) $\Rightarrow$ (i).

**Proof:** We shall start with (iii).

If $AGA = A$ then by Theorem 2.4, $C_r(A)C_r(G)C_r(A) = C_r(A)$. Thus (iii) $\Rightarrow$ (iv).

Now, let $C_r(A)$ be regular. Note that $C_r(A)$ is an $\binom{m}{r} \times \binom{n}{r}$ matrix of rank 1. Let $\{c_{\beta\alpha} : \beta \in Q_{r,m}, \alpha \in Q_{r,n}\}$ be an $\binom{n}{r} \times \binom{m}{r}$ matrix which is a $g$-inverse of $C_r(A) = \{|A_{\alpha\beta}| : \alpha \in Q_{r,m}, \beta \in Q_{r,n}\}$. From Theorem 5.1 we get that $|A_{\gamma\delta}|(\sum_{(\alpha,\beta)\in\Delta} |A_{\alpha\beta}|c_{\beta\alpha}) = |A_{\gamma\delta}|$ for all $(\gamma,\delta) \in \Delta$. Thus we have (iv) $\Rightarrow$ (v).

From (v) we get that $(\sum_{(\gamma,\delta)\in\Delta} |A_{\gamma\delta}|c_{\delta\gamma})(\sum_{(\alpha,\beta)\in\Delta} |A_{\alpha\beta}|c_{\beta\alpha}) = \sum_{(\gamma,\delta)\in\Delta} |A_{\gamma\delta}|c_{\delta\gamma}$ Thus $\sum_{(\alpha,\beta)\in\Delta} |A_{\alpha\beta}|c_{\beta\alpha}$ is idempotent. Since $\rho(A) = r$, there exists $(\gamma,\delta) \in \Delta$ such that $|A_{\gamma\delta}| \neq 0$. From (v) we get the $\sum |A_{\alpha\beta}|c_{\beta\alpha} \neq 0$. Now we shall look at the beginning of the string of implications.

(i) $\Rightarrow$ (ii) is clear.

We shall now look at (ii) $\Rightarrow$ (iii), the most crucial of all the implications.

Let $\{c_{\beta\alpha} : (\alpha,\beta) \in \Delta\}$ in $R$ be such that $(\sum_{(\alpha,\beta)\in\Delta} |A_{\alpha\beta}|c_{\beta\alpha})a_{k\ell} = a_{k\ell}$ for all $k$ and $\ell$.

Let us fix $\alpha = \{\alpha(1), \alpha(2), \cdots, \alpha(r)\}$ where $1 \le \alpha(1) < \alpha(2) < \cdots < \alpha(r) \le m$ and $\beta = \{\beta(1), \beta(2), \cdots, \beta(r)\}$ where $1 \le \beta(1) < \beta(2) < \cdots < \beta(r) \le n$ and indices $k$ and $\ell$.

Consider the matrix

$$B = \begin{bmatrix} & A_{\alpha\beta} & & a_{\alpha(1)\ell} \\ & & & a_{\alpha(2)\ell} \\ & & & \vdots \\ & & & a_{\alpha(r)\ell} \\ a_{k\beta(1)}, & a_{k\beta(2)}, & \cdots \quad a_{k\beta(r)} & a_{k\ell} \end{bmatrix}.$$

Since $\rho(A) = r$, irrespective of whether $k \in \alpha$ or $\ell \in \beta$ or not, $|B| = 0$. Hence using the result of section 2.3, we get

$$a_{k\ell}|A_{\alpha\beta}| = \sum_{i\in\alpha j\in\beta} a_{kj}a_{i\ell}\frac{\partial|A_{\alpha\beta}|}{\partial a_{ij}}$$

From this and the hypothesis of (ii) we get that

$$a_{k\ell} = a_{k\ell}\left(\sum_{(\alpha,\beta)\in\Delta}|A_{\alpha\beta}|c_{\beta\alpha}\right) = \sum_{(\alpha,\beta)\in\Delta}\left(\sum_{i\in\alpha j\in\beta} a_{kj}a_{i\ell}\frac{\partial|A_{\alpha\beta}|}{\partial a_{ij}}\right)c_{\beta\alpha}$$

By interchanging the summation signs we obtain,

$$a_{k\ell} = \sum_{i,j} a_{kj}a_{i\ell}\left[\sum_{\alpha:i\in\alpha}\sum_{\beta:j\in\beta}\frac{\partial|A_{\alpha\beta}|}{\partial a_{ij}}c_{\beta\alpha}\right].$$

If we call the quantity inside the square brackets as $g_{ji}$, then we get $a_{k\ell} = \sum_{i,j} a_{kj}g_{ji}a_{i\ell}$.

Thus $A = AGA$ where the $(j,i)$ element $g_{ji}$ of $G$ is given by

$$g_{ji} = \sum_{(\alpha,\beta)\in\Delta}\frac{\partial|A_{\alpha\beta}|}{\partial a_{ij}}c_{\beta\alpha}.$$

Thus A is regular. We have also manufactured a $g$-inverse of $A$ from a $\{c_{\beta\alpha} : (\alpha,\beta) \in \Delta\}$ satisfying the condition that $\left(\sum_{(\alpha,\beta)\in\Delta}|A_{\alpha\beta}|c_{\beta\alpha}\right)a_{k\ell} = a_{k\ell}$ for all $k$ and $\ell$. This completes the proof of all the implications.

Finally, if 0 and 1 are the only idempotents of $R$ the nonzero idempotent $\sum_{(\alpha,\beta)\in\Delta}|A_{\alpha\beta}|c_{\beta\alpha}$ is equal to 1. Hence (i) is satisfied.

♠

Indeed, the proof of the previous Theorem gives a method of constructing a $g$-inverse of a regular matrix if certain conditions are satisfied. This and other similar results are listed in the following Theorem.

**Theorem 5.4:** Let $A$ be a matrix over a commutative ring $R$ with $\rho(A) = r$.

(a) If $\{c_{\beta\alpha} : (\alpha, \beta) \in \Delta\}$ in $R$ are such that $(\sum_{(\alpha,\beta)\in\Delta} |A_{\alpha\beta}|c_{\beta\alpha})a_{k\ell} = a_{k\ell}$ for all $k$ and $\ell$ then $G = (g_{ij})$ defined by

$$g_{ij} = \sum_{(\alpha,\beta)\in\Delta} \frac{\partial|A_{\alpha\beta}|}{\partial a_{ji}} c_{\beta\alpha}$$

is a $g$-inverse of $A$.

(b) If $\{c_{\beta\alpha} : \alpha, \beta \in \Delta\}$ in $R$ are such that $G = (g_{ij})$ defined by

$$g_{ij} = \sum_{(\alpha,\beta)\in\Delta} \frac{\partial|A_{\alpha\beta}|}{\partial a_{ji}} c_{\beta\alpha}$$

is a $g$-inverse of $A$ then

$$\left(\sum_{(\alpha,\beta)\in\Delta} |A_{\alpha\beta}|c_{\beta\alpha}\right) a_{k\ell} = a_{k\ell}$$

for all $k$ and $\ell$.

**Proof:** The proof of (a) is already available in the proof of Theorem 5.3.

For the proof of (b) let $G = (g_{ij})$ where

$$g_{ij} = \sum_{(\alpha,\beta)\in\Delta} \frac{\partial|A_{\alpha\beta}|}{\partial a_{ji}} c_{\beta\alpha}$$

and let $AGA = A$. This gives us that $\sum_{j,i} a_{kj}g_{ji}a_{i\ell} = a_{k\ell}$ for all $k$ and $\ell$. i.e.,

$$\sum_{j,i} a_{kj} \left(\sum_{(\alpha,\beta)\in\Delta} \frac{\partial|A_{\alpha\beta}|}{\partial a_{ij}} c_{\beta\alpha}\right) a_{i\ell} = a_{k\ell}$$

for all $k$ and $\ell$.

This can be rewritten as

$$\sum_{(\alpha,\beta)\in\Delta} \left(\sum_{i\in\alpha j\in\beta} a_{kj} \frac{\partial|A_{\alpha\beta}|}{\partial a_{ij}} a_{i\ell}\right) c_{\beta\alpha} = a_{k\ell}.$$

But from the result of section 2.3 we have,

$$a_{k\ell}|A_{\alpha\beta}| = \sum_{i\in\alpha j\in\beta} a_{kj}a_{i\ell}\frac{\partial|A_{\alpha\beta}|}{\partial a_{ij}}$$

for all $k, \ell, \alpha$ and $\beta$.

Hence $a_{k\ell}(\sum_{(\alpha,\beta)\in\Delta} |A_{\alpha\beta}|c_{\beta\alpha}) = a_{k\ell}$ for all $k$ and $\ell$.

♠

Let us give a quick example to show that for the statements in Theorem 5.3 (iv) ⇒ (iii) is not true in general for commutative rings.

**Example:** Consider the ring $\mathbb{Z}[12]$. Let $A = \begin{bmatrix} 2 & 0 \\ 0 & 2 \end{bmatrix}$.

For the element 2 of $\mathbb{Z}[12]$ there is no $g \in \mathbb{Z}[12]$ such that $2 \cdot g \cdot 2 = 2$. Hence $\begin{bmatrix} 2 & 0 \\ 0 & 2 \end{bmatrix}$ is not a regular matrix. However, $\rho(A) = 2$ and $C_2(A) = 4$. Also $4 \cdot 4 \cdot 4 = 4$ in $\mathbb{Z}[12]$. Thus $C_2(A)$ is regular but $A$ is not regular. Note that $\mathbb{Z}[12]$ has zero divisors.

## 5.3 A result of Prasad and Robinson

In the previous section (and also in section 3.4) we had constructed a matrix $G$ of order $n \times m$ starting with a matrix $(c_{\beta\alpha})_{\alpha\in Q_{r,m},\beta\in Q_{r,n}}$ and showed that $AGA = A$ holds if certain conditions are satisfied. K.M. Prasad and D.W. Robinson have analyzed this construction in detail for matrices over commutative rings. We shall present these results as a comprehensive Theorem.

Let $R$ be a commutative ring. We shall consider matrices over $R$. Let $A$ be an $m \times n$ matrix over $R$ with $\rho(A) = r$. Let $S = (s_{\beta\alpha})_{\alpha\in Q_{r,m},\beta\in Q_{r,n}}$ be an $\binom{n}{r} \times \binom{m}{r}$ matrix. Define an $n \times m$ matrix $G = (g_{ij})$ by

$$g_{ij} = \sum_{\alpha\in Q_{r,m},\beta\in Q_{r,n}} \frac{\partial|A_{\alpha\beta}|}{\partial a_{ji}} s_{\beta\alpha}.$$

Since $G$ depends on $S$ we shall denote $G$ by $A_S$. $A_S$ can also be represented as $\sum_{\alpha\in Q_{r,m},\beta\in Q_{r,n}} s_{\beta\alpha}P_{\alpha\beta}$ where $P_{\alpha\beta}$ is the $n\times m$ matrix $\begin{bmatrix} 0 & 0 & 0 \\ 0 & A^{adj}_{\alpha\beta} & 0 \\ 0 & 0 & 0 \end{bmatrix}$ where $A^{adj}_{\alpha\beta}$ is the *adjoint* of the matrix $A_{\alpha\beta}$ distributed in the $\beta$ row indices and $\alpha$ column indices of $P_{\alpha\beta}$. We do not need to use this representation, but note that if $A$ is a unimodular $n \times n$ matrix then this representation turns out to be the inverse obtained by using the Cramer Rule.

Thus, for any given matrix $A$ with $\rho(A) = r$ we have a correspondence $S \to A_S$ from $\binom{n}{r} \times \binom{m}{r}$ matrices to $n \times m$ matrices. The relation of $S$ to $C_r(A)$ is reflected in the relation of $A_S$ to $A$.

**Theorem 5.5 (Prasad and Robinson):** Let $A$ be an $m \times n$ matrix with $\rho(A) = r$ and $S$ be an $\binom{n}{r} \times \binom{m}{r}$ matrix over a commutative ring $R$ with an involution $a \to \bar{a}$. Then

(a) $AA_SA = Tr(C_r(A)S)A$.

(b) If $\rho(S) = 1, A_SAA_S = Tr(C_r(A)S)A_S$.

(c) If $(C_r(A)S)^* = C_r(A)S$ then $(AA_S)^* = AA_S$.

(d) If $(SC_r(A))^* = SC_r(A)$ then $(A_SA)^* = A_SA$.

(e) If $C_r(A)S = SC_r(A)$ then $AA_S = A_SA$.

**Proof:** (a) was essentially proved in Theorem 5.3. In fact, letting $A_S = G$,

$$\begin{aligned}
(AA_SA)_{k\ell} &= (AGA)_{k\ell} = \sum_{i,j} a_{kj} g_{ji} a_{i\ell} \\
&= \sum_{i,j} a_{kj} a_{i\ell} \sum_{\alpha,\beta} \frac{\partial |A_{\alpha\beta}|}{\partial a_{ij}} s_{\beta\alpha} \\
&= \sum_{i,j} a_{kj} a_{i\ell} \sum_{\substack{\alpha : i \in \alpha \\ \beta : j \in \beta}} \frac{\partial |A_{\alpha\beta}|}{\partial a_{ij}} s_{\beta\alpha} \\
&= \sum_{\alpha,\beta} \sum_{\substack{i : i \in \alpha \\ j : j \in \beta}} a_{kj} a_{i\ell} \frac{\partial |A_{\alpha\beta}|}{\partial a_{ij}} s_{\beta\alpha} \\
&= a_{k\ell} \sum_{\substack{\alpha \in Q_{r,m} \\ \beta \in Q_{r,n}}} |A_{\alpha\beta}| s_{\beta\alpha} \quad \text{(as in the proof of Theorem 5.3)} \\
&= Tr(C_r(A)S)(A)_{k\ell}.
\end{aligned}$$

(b) Let us look at $(A_SAA_S)_{k\ell}$.

$$\begin{aligned}
(A_SAA_S)_{k\ell} &= (GAG)_{k\ell} = \sum_{i,j} g_{kj} a_{ji} g_{i\ell} \\
&= \sum_{i,j} \left( \sum_{\alpha,\beta} \frac{\partial |A_{\alpha\beta}|}{\partial a_{jk}} s_{\beta\alpha} \right) a_{ji} \left( \sum_{\gamma,\delta} \frac{\partial |A_{\gamma\delta}|}{\partial a_{\ell i}} s_{\delta\gamma} \right) \\
&= \sum_{\alpha,\beta,\gamma,\delta} \left( \sum_{i,j} \frac{\partial |A_{\alpha\beta}|}{\partial a_{jk}} a_{ji} \frac{\partial |A_{\gamma\delta}|}{\partial a_{\ell i}} \right) s_{\beta\alpha} s_{\delta\gamma}.
\end{aligned}$$

Let us evaluate

$$\sum_{\substack{i,j \\ j\in\alpha, i\in\delta}} \frac{\partial|A_{\alpha\beta}|}{\partial a_{jk}} a_{ji} \frac{\partial|A_{\gamma\delta}|}{\partial a_{\ell i}}$$

for fixed $\alpha, \beta, \gamma$ and $\delta$. If $k \notin \beta$ or $\ell \notin \gamma$ this quantity equals 0.

If $k \in \beta$ and $\ell \in \gamma$, consider the matrix

$$T = \begin{bmatrix} A_{\gamma-\{\ell\},\beta-\{k\}} & A_{\gamma-\{\ell\},\delta} \\ A_{\alpha,\beta-\{k\}} & A_{\alpha,\delta} \end{bmatrix}.$$

Then $T$ is an $((r-1)+r) \times ((r-1)+r)$ matrix and $|T| = 0$.

An application of Proposition 2.6 to $T$ and simplification gives us that

$$\begin{vmatrix} A_{\gamma-\{\ell\},\beta-\{k\}} & 0 \\ A_{\alpha,\beta-\{k\}} & A_{\alpha,\delta} \end{vmatrix} = (-1)^{s-1} \begin{vmatrix} 0 & A_{\gamma-\{\ell\},\delta} \\ A_{\alpha,\beta-\{k\}} & A_{\alpha,\delta} \end{vmatrix}$$

Expanding these determinants we get,

$$\sum_{\substack{i,j \\ j\in\alpha, i\in\delta}} \frac{\partial|A_{\alpha\beta}|}{\partial a_{jk}} a_{ji} \frac{\partial|A_{\gamma\delta}|}{\partial a_{\ell i}} = |A_{\alpha\delta}| \frac{\partial|A_{\gamma\beta}|}{\partial a_{\ell k}}$$

Also, since $\rho(S) = 1$, $s_{\beta\alpha}s_{\delta\gamma} = s_{\beta\gamma}s_{\delta\alpha}$.

$$\begin{aligned} \text{Hence } (A_S A A_S)_{k\ell} &= \sum_{\alpha,\beta,\gamma,\delta} |A_{\alpha\delta}| \frac{\partial|A_{\gamma\beta}|}{\partial a_{\ell k}} s_{\beta\gamma}s_{\delta\alpha} \\ &= \left(\sum_{\alpha,\delta} |A_{\alpha\delta}| s_{\delta\alpha}\right) \left(\sum_{\beta,\gamma} \frac{\partial|A_{\gamma\beta}|}{\partial a_{\ell k}} s_{\beta\gamma}\right) \\ &= Tr(C_r(A)S)(A_S)_{k\ell}. \end{aligned}$$

(c) Let us find $(AA_S)_{k\ell}$.

$$\begin{aligned} (AA_S)_{k\ell} &= \sum_i a_{ki} g_{i\ell} \\ &= \sum_i a_{ki} \sum_{\substack{\alpha\in Q_{r,m} \\ \beta\in Q_{r,n}}} \frac{\partial|A_{\alpha\beta}}{\partial a_{\ell i}}| s_{\beta\alpha} \\ &= \sum_{\substack{\alpha\in Q_{r,m} \\ \beta\in Q_{r,n}}} \left(\sum_{i:i\in\beta} a_{ki} \frac{\partial|A_{\alpha\beta}|}{\partial a_{\ell i}}\right) s_{\beta\alpha}. \end{aligned}$$

But, for fixed $\alpha, \beta, k$ and $\ell$,

$$\begin{aligned}\sum_{i:i\in\beta} a_{ki}\frac{\partial}{\partial a_{\ell i}}|A_{\alpha\beta}| &= 0 \text{ if } \ell \notin \alpha \\ &= |A_{\alpha\beta}| \text{ if } \ell \in \alpha \text{ and } \ell = k \\ &= 0 \text{ if } \ell \in \alpha, \ell \neq k \text{ and } k \in \alpha \\ &= |A_{\alpha-\{\ell\}\cup\{k\},\beta}| \text{ if } \ell \in \alpha \text{ and } k \notin \alpha.\end{aligned}$$

Hence, for $k = \ell$,

$$\begin{aligned}(AA_S)_{\ell\ell} &= \sum_{\substack{\alpha\in Q_{r,m}\\ \ell\in\alpha}} \sum_{\beta\in Q_{r,n}} |A_{\alpha\beta}|s_{\beta\alpha} \\ &= \sum_{\substack{\alpha\in Q_{r,m}\\ \ell\in\alpha}} (C_r(A)S)_{\alpha\alpha}.\end{aligned}$$

For $k \neq \ell$,

$$\begin{aligned}(AA_S)_{k\ell} &= \sum_{\substack{\alpha\in Q_{r,m}\\ \ell\in\alpha\\ k\notin\alpha}} \sum_{\beta\in Q_{r,n}} |A_{\alpha-\{\ell\}\cup\{k\},\beta}|s_{\beta\alpha} \\ &= \sum_{\substack{\alpha\in Q_{r,m}\\ \ell\in\alpha\\ k\notin\alpha}} (C_r(A)S)_{\alpha-\{\ell\}\cup\{k\},\alpha} \\ &= \sum_{\substack{\gamma\in Q_{r-1,m}\\ \ell\notin\gamma\\ k\notin\gamma}} (C_r(A)S)_{\gamma\cup\{k\},\gamma\cup\{\ell\}}.\end{aligned}$$

Now, if $(C_r(A)S)^* = C_r(A)S$, then for $k = \ell$,

$$\begin{aligned}((AA_S)^*)_{\ell\ell} &= \sum_{\substack{\alpha\in Q_{r,m}\\ \ell\in\alpha}} \overline{(C_r(A)S)}_{\alpha\alpha} \\ &= \sum_{\alpha\in Q_{r,m}} (C_r(A)S)_{\alpha\alpha} = (AA_S)_{\ell\ell}.\end{aligned}$$

For $k \neq \ell$,

$$\begin{aligned}((AA_S)^*)_{k\ell} &= \overline{(AA_S)}_{\ell k} \\ &= \sum_{\substack{\gamma\in Q_{r-1,m}\\ \ell\notin\gamma\\ k\notin\gamma}} \overline{(C_r(A)S)}_{\gamma\cup\{k\},\gamma\cup\{\ell\}}\end{aligned}$$

$$= \sum_{\substack{\gamma \in Q_{r-1,m} \\ \ell \notin \gamma \\ k \notin \gamma}} (C_r(A)S)_{\gamma \cup \{\ell\}, \gamma \cup \{k\}}$$
$$= (AA_S)_{k\ell}.$$

Thus $(AA_S)^* = AA_S$.

(d) is proved similarly.

(e) We have already evaluated $(AA_S)_{k\ell}$ in the proof of (c). Let us look at $(A_SA)_{k\ell}$.

$$\begin{aligned}
(A_SA)_{k\ell} = (GA)_{k\ell} &= \sum_i g_{ki}a_{i\ell} \\
&= \sum_i \sum_{\substack{\alpha \in Q_{r,m} \\ \beta \in Q_{r,n}}} \frac{\partial |A_{\alpha\beta}|}{\partial a_{ik}} s_{\beta\alpha} a_{i\ell} \\
&= \sum_{\substack{\alpha \in Q_{r,m} \\ \beta \in Q_{r,n}}} \left( \sum_{i:i\in\alpha} a_{i\ell} \frac{\partial |A_{\alpha\beta}|}{\partial a_{ik}} \right) s_{\beta\alpha}.
\end{aligned}$$

But, for fixed $\alpha, \beta, k$ and $\ell$,

$$\begin{aligned}
\sum_{i:i\in\alpha} a_{i\ell} \frac{\partial |A_{\alpha\beta}|}{\partial a_{ik}} &= 0 \text{ if } k \notin \beta \\
&= |A_{\alpha\beta}| \text{ if } k \in \beta \text{ and } \ell = k \\
&= 0 \text{ if } k \in \beta, \ell \neq k \text{ and } \ell \in \beta \\
&= |A_{\alpha,\beta-\{k\}\cup\{\ell\}}| \text{ if } k \in \beta, \ell \neq k \text{ and } \ell \notin \beta.
\end{aligned}$$

Hence, for $k = \ell$,

$$\begin{aligned}
(A_SA)_{\ell\ell} &= \sum_{\substack{\beta \in Q_{r,n} \\ \ell \in \beta}} \sum_{\alpha} |A_{\alpha\beta}| s_{\beta\alpha} \\
&= \sum_{\substack{\beta \in Q_{r,n} \\ \ell \in \beta}} (SC_r(A))_{\beta\beta}.
\end{aligned}$$

For $k \neq \ell$,

$$(A_SA)_{k\ell} = \sum_{\substack{\beta \in Q_{r,n} \\ k \in \beta \\ \ell \notin \beta}} \sum_{\alpha \in Q_{r,m}} |A_{\alpha,\beta-|\{k\}\cup\{\ell\}}| s_{\beta\alpha}$$

$$= \sum_{\substack{\beta \in Q_{r,n} \\ k \in \beta \\ \ell \notin \beta}} (SC_r(A))_{\beta, \beta - \{k\} \cup \{\ell\}}$$

$$= \sum_{\substack{\gamma \in Q_{r-1,n} \\ k \notin \gamma \\ \ell \notin \gamma}} (SC_r(A))_{\gamma \cup \{k\}, \gamma \cup \{\ell\}}.$$

If $C_r(A)S = SC_r(A)$ then clearly $(AA_S)_{k\ell} = (A_SA)_{k\ell}$ for every $k$ and $\ell$. ♠

We shall go to integral domains in the next Chapter.

**Proof:** Let $F$ be the field of quotients of the integral domain $R$. The matrix $A$, which is a matrix over $F$ also, has a rank factorization over $F$. Let $A = BC$ where $B$ is an $m \times r$ matrix, and $C$ is an $r \times n$ matrix, be a rank factorization of $A$ over $F$.

Since $G$ is a reflexive $g$-inverse of $A$ we get that $AGA = BCGBC = BC$ and $GBCG = G$. Since $C$ has a right inverse and $B$ has a left inverse over $F$ we get that $CGB = I$. Let us write $M = GB$ and $N = CG$. Then, it can be easily seen that $G = MN$ is a rank factorization of $G$ over $F$. Let $M = (m_{ij})$ and $N = (n_{ij})$. Also, for $\alpha \in Q_{r,m}$ and $\beta \in Q_{r,n}$, from the Cauchy-Binet Theorem and the results of section 2.3 we have that $|G_{\beta\alpha}| = |M_{\beta *}||N_{*\alpha}|$ and

$$\frac{\partial |A_{\alpha\beta}|}{\partial a_{ji}} = \sum_{k=1}^{r} \frac{\partial |B_{\alpha *}|}{\partial b_{jk}} \frac{\partial |C_{*\beta}|}{\partial c_{ki}}$$

for every $i, j$.

Now,

$$\begin{aligned}
\sum_{(\alpha,\beta)\in\Delta} \frac{\partial |A_{\alpha\beta}|}{\partial a_{ji}} |G_{\beta\alpha}| &= \sum_{\alpha: j\in\alpha\in Q_{r,m}} \sum_{\beta: i\in\beta\in Q_{r,n}} \frac{\partial |A_{\alpha\beta}|}{\partial a_{ji}} |G_{\beta\alpha}| \\
&= \sum_{\alpha: j\in\alpha\in Q_{r,m}} \sum_{\beta: i\in\beta\in Q_{r,n}} \sum_{k=1}^{r} \frac{\partial |B_{\alpha *}|}{\partial b_{jk}} \frac{\partial C_{*\beta}|}{\partial c_{ki}} ||M_{\beta *}||N_{*\alpha}| \\
&= \sum_{k=1}^{r} \left( \sum_{\alpha: j\in\alpha\in Q_{r,m}} \frac{\partial |B_{\alpha *}|}{\partial b_{jk}} |N_{*\alpha}| \right) \left( \sum_{\beta: i\in\beta\in Q_{r,n}} \frac{\partial |C_{*\beta}|}{\partial c_{ki}} |M_{\beta *}| \right).
\end{aligned}$$

But since $CM = I$ and $NB = I$, by Theorem 3.8 we get that

$$\sum_{\beta: i\in\beta\in Q_{r,n}} \frac{\partial |C_{*\beta}|}{\partial c_{ki}} |M_{\beta *}| = \sum_{\beta\in Q_{r,n}} \frac{\partial |C_{*\beta}|}{\partial c_{ki}} |M_{\beta *}| = m_{ik}.$$

Also,

$$\sum_{\alpha: j\in\alpha\in Q_{r,m}} \frac{\partial |B_{\alpha *}|}{\partial b_{jk}} |N_{*\alpha}| = n_{kj}.$$

Hence,

$$\sum_{(\alpha,\beta)\in\Delta} \frac{\partial |A_{\alpha\beta}|}{\partial a_{ji}} |G_{\beta\alpha}| = \sum_{k=1}^{r} m_{ik} n_{kj} = g_{ij}.$$

♠

Thus, we have shown that, if $G = (g_{ij})$ is a reflexive $g$-inverse of $A$, then there exists $(c_{\beta\alpha})_{(\alpha,\beta)\in\Delta} \left(= (|G_{\beta\alpha}|)_{(\alpha,\beta)\in\Delta}\right)$ such that $\sum_{(\alpha,\beta)\in\Delta} |A_{\alpha\beta}|c_{\beta\alpha} = 1$ and

$$g_{ij} = \sum_{(\alpha,\beta)\in\Delta} \frac{\partial|A_{\alpha\beta}|}{\partial a_{ji}} c_{\beta\alpha}.$$

If we are given a general $C = (c_{\beta\alpha})_{(\alpha,\beta)\in\Delta}$, an $\binom{n}{r} \times \binom{m}{r}$ matrix such that $\sum_{(\alpha,\beta)\in\Delta} |A_{\alpha\beta}|c_{\beta\alpha} = 1$ then, what are the conditions on $C$ so that $G$ defined above gives a reflexive $g$-inverse of $A$? We shall deal with this in the next Theorem. A more general result than the one proved for integral domains here will be proved later in section 7.5.

**Theorem 6.3:** Let $A$ be an $m \times n$ matrix over an integral domain $R$ with $\rho(A) = r$. Let $C = (c_{\beta\alpha})_{(\alpha,\beta)\in\Delta}$ be an $\binom{n}{r} \times \binom{m}{r}$ matrix over $R$ such that $\sum_{(\alpha,\beta)\in\Delta} |A_{\alpha\beta}|c_{\beta\alpha} = 1$. Let $G = (g_{ij})$ be defined by $g_{ij} = \sum_{(\alpha,\beta)\in\Delta} \frac{\partial|A_{\alpha\beta}|}{\partial a_{ji}} c_{\beta\alpha}$ for all $i, j$.

If $\rho(C) = 1$ then $G$ is a reflexive $g$-inverse of $A$.

Conversely every reflexive $g$-inverse of $A$ can be constructed from a $C = (c_{\beta\alpha})_{(\alpha,\beta)\in\Delta}$ with $\rho(C) = 1$ and $\sum_{(\alpha,\beta)\in\Delta} |A_{\alpha\beta}|c_{\beta\alpha} = 1$ using the above procedure.

**Proof:** We have already seen that $G$ is a $g$-inverse of $A$. Hence $\rho(G) \geq r$. To show that $G$ is a reflexive $g$-inverse of $A$ by Theorem 3.4 (d) it is enough to show that $\rho(G) \leq r$.

Let $F$ be the field of quotients of $R$. We shall show that the $n \times m$ matrix $G$ can be written as $MN$ where $M$ is an $n \times r$ matrix over $F$ and $N$ is an $r \times m$ matrix over $F$. This will imply that $\rho(G) \leq r$.

Let $A = BH$ be a rank factorization of $A$ over the field $F$. Let $B = (b_{ij})$ and $H = (h_{ij})$. Also write $C = DE$ over the field $F$ where $D$ is an $\binom{n}{r} \times 1$ matrix and $E$ is a $1 \times \binom{m}{r}$ matrix. Let $D = (d_{\beta 1})_{\beta\in Q_{r,n}}$ and $E = (e_{1\alpha})_{\alpha\in Q_{r,m}}$. Then $c_{\beta\alpha} = d_{\beta 1}e_{1\alpha}$. Recalling from section 2.3 that

$$\frac{\partial|A_{\alpha\beta}|}{\partial a_{ji}} = \sum_{k=1}^{r} \frac{\partial|B_{\alpha *}|}{\partial b_{jk}} \frac{\partial|H_{*\beta}|}{\partial h_{ki}}$$

we have,

$$g_{ij} = \sum_{(\alpha,\beta)\in\Delta} c_{\beta\alpha} \frac{\partial|A_{\alpha\beta}|}{\partial a_{ji}}$$

$$= \sum_{(\alpha,\beta)\in\Delta} d_{\beta 1} e_{1\alpha} \sum_{k=1}^{r} \frac{\partial |B_{\alpha *}|}{\partial b_{jk}} \frac{\partial |H_{*\beta}|}{\partial h_{ki}}$$

$$= \sum_{k=1}^{r} \left( \sum_{\beta\in Q_{r,n}} d_{\beta 1} \frac{\partial |H_{*\beta}|}{\partial h_{ki}} \right) \left( \sum_{\alpha\in Q_{r,m}} e_{1\alpha} \frac{\partial |B_{\alpha *}|}{\partial b_{jk}} \right)$$

$$= \sum_{k=1}^{r} m_{ik} n_{kj}$$

where $m_{ik} = \sum_{\beta\in Q_{r,n}} d_{\beta 1} \frac{\partial |H_{*\beta}|}{\partial h_{ki}}$ and $n_{kj} = \sum_{\alpha\in Q_{r,m}} e_{1\alpha} \frac{\partial |B_{\alpha *}|}{\partial b_{jk}}$.

If we write $M = (m_{ik})$ and $N = (n_{kj})$ then $M$ is an $n \times r$ matrix over $F$, $N$ is an $r \times m$ matrix over $F$ and $G = MN$. Thus $\rho(G) = r$.

The second part of the Theorem is clear from Theorem 6.2.

♠

## 6.3 M-P inverses over integral domains

In this section we shall find necessary and sufficient conditions for a matrix over an integral domain with an involution $a \to \bar{a}$ to admit a Moore-Penrose inverse. We shall also give a construction of the Moore-Penrose inverse when it exists. We shall look at full column rank matrices, matrices with rank factorizations, matrices of rank one and general $m \times n$ matrices in that order.

Let $R$ be an integral domain with an involution $a \to \bar{a}$. Recall that for an $m \times n$ matrix $A, A^* = \bar{A}^T$. Note also that for $\alpha \in Q_{k,n}$ and $\beta \in Q_{k,m}$, $|A^*_{\alpha\beta}| = |\bar{A}^T_{\alpha\beta}| = |\bar{A}_{\beta\alpha}| = \overline{|A_{\beta\alpha}|}$.

**Proposition 6.4:** Let $L$ be an $m \times n$ matrix over an integral domain $R$ with an involution $a \to \bar{a}$. Let $\rho(L) = n$. Then $L$ has Moore-Penrose inverse if and only if $L^*L$ is unimodular. In fact, if $G$ is the Moore-Penrose inverse of $L$ then $(GG^*)(L^*L) = I$ (and also of course, $(L^*L)(GG^*) = I$). If $L^*L$ is unimodular, then $(L^*L)^{-1}L^*$ is the Moore-Penrose inverse of $L$. A similar result also holds for matrices with full row rank.

**Proof:** Let $F$ be the field of quotients of $R$. Let $G$ be the Moore-Penrose inverse of $L$. Since $L$ has a left inverse over $F$, $LGL = L$ gives us that $GL = I$. $(LG)^* = LG$ together with $LGL = L$ gives us that $G^*L^*L = L$. Hence $GG^*L^*L = GL = I$. Thus $L^*L$ is unimodular.

If $L^*L$ is unimodular let us show that $G = (L^*L)^{-1}L^*$ is the Moore-Penrose inverse of $L$. Clearly $GL = I$. Hence, $LGL = L$ and $GLG = G$.

Since $((L^*L)^{-1})^* = (L^*L)^{-1}$ we get that $(LG)^* = LG$. Thus $G$ is the Moore-Penrose inverse of $L$.

♠

From this Proposition we can quickly derive necessary and sufficient conditions for a matrix with a rank factorization to have Moore-Penrose inverse.

**Proposition 6.5:** Let $A$ be an $m \times n$ matrix over an integral domain $R$ with an involution $a \to \bar{a}$. Let $\rho(A) = r$. Also let $A = LR$ where $L$ is an $m \times r$ matrix and $R$ is an $r \times n$ matrix over the integral domain (we are not asking that $A = LR$ is a rank factorization). Then the following are equivalent.

(i) $A^+$ exists.

(ii) $L^+$ and $R^+$ exist.

(iii) $L^*L$ and $RR^*$ are invertible.

(iv) $\sum_{\alpha \in Q_{r,m}} \overline{|L_{\alpha *}|}|L_{\alpha *}|$ and $\sum_{\beta \in Q_{r,n}} \overline{|R_{*\beta}|}|R_{*\beta}|$ are units of the integral domain.

(v) $\sum_{(\alpha,\beta) \in \Delta} \overline{|A_{\alpha\beta}|}|A_{\alpha\beta}|$ is a unit of $R$.

Further, $L^+ = RA^+, R^+ = A^+L$ and $A^+ = R^+L^+$.

**Proof:** (i) $\Rightarrow$ (ii). By considering the field $F$ of quotients of $R$ one can easily verify that $L^+ = RA^+$ and $R^+ = A^+L$ will serve the purpose.

(ii) $\Rightarrow$ (iii) follows from Proposition 6.4.

(iii) $\Rightarrow$ (iv). Since $L^*L$ is unimodular, by Theorem 3.6 we have that $|L^*L|$ is a unit of the integral domain. But by the Cauchy-Binet Theorem (Theorem 2.2), $|L^*L| = \sum_{\alpha \in Q_{r,m}} \overline{|L_{\alpha *}|}|L_{\alpha *}|$. Hence $\sum_{\alpha \in Q_{r,m}} \overline{|L_{\alpha *}|}|L_{\alpha *}|$ is a unit of the integral domain. Similarly $\sum_{\beta \in Q_{r,n}} \overline{|R_{*\beta}|}|R_{*\beta}|$ is also a unit of the integral domain.

(iv) $\Rightarrow$ (v). Since $A = LR$, $L$ is an $m \times r$ matrix and $R$ is an $r \times n$ matrix, we get that, for $\alpha \in Q_{r,m}$ and $\beta \in Q_{r,n}$ $|A_{\alpha\beta}| = |L_{\alpha *}||R_{*\beta}|$. Hence

$$\sum_{(\alpha,\beta) \in \Delta} \overline{|A_{\alpha\beta}|}|A_{\alpha\beta}| = \left( \sum_{\alpha \in Q_{r,m}} \overline{|L_{\alpha *}|}|L_{\alpha *}| \right) \left( \sum_{\beta \in Q_{r,n}} \overline{|R_{*\beta}|}|R_{*\beta}| \right).$$

Since $\sum_{\alpha \in Q_{r,m}} \overline{|L_{\alpha *}|}|L_{\alpha *}|$ and $\sum_{\beta \in Q_{r,n}} \overline{|R_{*\beta}|}|R_{*\beta}|$ both are units of the integral domain we get that $\sum_{(\alpha,\beta) \in \Delta} \overline{|A_{\alpha\beta}|}|A_{\alpha\beta}|$ is a unit of the integral domain.

We shall now retrace all the steps.

(v) $\Rightarrow$ (iv) is clear because if $a = bc$ in a commutative ring and $a$ is a unit then both $b$ and $c$ are units in $R$.

(iv) $\Rightarrow$ (iii) is clear because $|L^*L| = \overline{\sum_{\alpha \in Q_{r,m}} |L_{\alpha *}|}|L_{\alpha *}|$ is a unit of the integral domain. By Theorem 3.6 $L^*L$ is invertible. Similarly $RR^*$ is also invertible.

(iii) $\Rightarrow$ (ii) follows from Proposition 6.4.

(ii) $\Rightarrow$ (i). One easily verifies that $R^+L^+$ is the Moore-Penrose inverse of $A = LR$.

♠

However, we have observed earlier in section 3.4 that over an integral domain not every matrix need have rank factorization. We are going to show that (i) and (v) of the previous Proposition are equivalent for every matrix over an integral domain with an involution $a \to \bar{a}$, whether the matrix has a rank factorization or not. Towards this, let us first look at rank one matrices.

**Theorem 6.6:** Let $A = (a_{ij})$ be an $m \times n$ matrix over an integral domain with an involution $a \to \bar{a}$. Let $\rho(A) = 1$. Then the following are equivalent.

(i) $A^+$ exists.

(ii) $\sum_{i,j} \overline{a_{ij}} a_{ij}$ is a unit.

In fact, if $A^+ = G = (g_{ij})$ then $\left(\sum_{i,j} \overline{a_{ij}} a_{ij}\right)\left(\sum_{i,j} \overline{g_{ij}} g_{ij}\right) = 1$ and if $\sum_{i,j} \overline{a_{ij}} a_{ij} = Tr(A^*A) = u$ is a unit then $A^+ = u^{-1}A^*$.

**Proof:** (i) $\Rightarrow$ (ii). Let $F$ be the field of quotients of the integral domain. The involution $a \to \bar{a}$ also extends in a natural way to $F$. Even though $A$ may not have a rank factorization over the integral domain, $A$ definitely has a rank factorization $A = LR$ over $F$. $L$ is an $m \times 1$ matrix and $R$ is a $1 \times n$ matrix.

Let $A^+ = G = (g_{ij})$. Since $G$ is a reflexive $g$-inverse of $A, \rho(G) = \rho(A) = 1$ by Theorem 3.4. Let $G = KU$ be a rank factorization of $G$ over $F$. First of all, note that $AGA = A$ gives us that $LRKULR = LR$. Since $L$ has a left inverse and $R$ has a right inverse we get that $(RK)(UL) = 1$.

Since $A^+ = G$, by Proposition 6.5 we get that $L^+ = RG = RKU$ and $R^+ = GL = KUL$. By Proposition 6.4 we get that

$$(L^*L)((RK)UU^*(RK)^*) = 1 \quad \text{and} \quad (RR^*)(UL)^*K^*K(UL) = 1.$$

But $RK$ and $UL$ both are $1 \times 1$ matrices and $(RK)(UL) = 1$.

Hence $(L^*L)(UU^*)(RK)^*(RK)(RR^*)(K^*K)(UL)^*(UL) = (L^*L)(UU^*)(RR^*)(K^*K) = 1$

Since $A = LR, \sum_{i,j} \overline{a_{ij}} a_{ij} = Tr(A^*A) = Tr(R^*L^*LR) = Tr(L^*LRR^*) = (L^*L)(RR^*)$. Similarly $Tr(G^*G) = (K^*K)(UU^*)$.

Thus, $Tr(G^*G)Tr(A^*A) = 1$. Thus

$$\left(\sum_{i,j} \overline{g_{ij}} g_{ij}\right)\left(\sum_{i,j} \overline{a_{ij}} a_{ij}\right) = 1.$$

To prove (ii) $\Rightarrow$ (i), we shall show that $u^{-1}A^*$ is the Moore-Penrose inverse of $A$ where $u = \sum_{i,j} \overline{a_{ij}} a_{ij}$, is a unit of $R$. Let $G = u^{-1}A^*$.

Clearly $(AG)^* = AG$ and $(GA)^* = GA$. Now, for any $i, j, (AGA)_{ij} = u^{-1}(AA^*A)_{ij} = u^{-1} \sum a_{ik} \overline{a_{\ell k}} a_{\ell j}$. But since $\rho(A) = 1$ we have $a_{ik} a_{\ell j} = a_{ij} a_{\ell k}$ for all $i, j, k$ and $\ell$.

Hence

$$\begin{aligned}(AGA)_{ij} &= u^{-1} \sum_{k,\ell} a_{ij} \overline{a_{\ell k}} a_{\ell k} \\ &= u^{-1} a_{ij} u = a_{ij} = (A)_{ij}.\end{aligned}$$

Thus $AGA = A$. Also,

$$(GAG)_{ij} = u^{-2}(A^*AA^*)_{ij} = u^{-2} \sum_{k\ell} \overline{a_{ki}} a_{k\ell} \overline{a_{j\ell}}.$$

But since $\rho(\bar{A}) = \rho(A) = 1$ we have $\overline{a_{ki}} \overline{a_{j\ell}} = \overline{a_{k\ell}} \overline{a_{ji}}$ for all $i, j, k$ and $\ell$. Hence, $(GAG)_{ij} = u^{-2} \sum_{k,\ell} \overline{a_{ji}} \overline{a_{k\ell}} a_{k\ell} = u^{-1} \overline{a_{ji}} = u^{-1}(A^*)_{ij} = (G)_{ij}$. Thus $GAG = G$.

Thus $u^{-1}A^*$ is the Moore-Penrose inverse of $A$.

♠

Since, for a matrix $A$ with $\rho(A) = 1$, the rank of the $r$th compound matrix is equal to one, the question of finding necessary and sufficient conditions for the existence of $A^+$ can be transferred to the corresponding question for $C_r(A)$. The next Theorem also gives a method of constructing $A^+$ when it exists.

**Theorem 6.7:** Let $A$ be a $m \times n$ matrix over an integral domain $R$ with an involution $a \to \bar{a}$. Then the following are equivalent.

(i) $A^+$ exists.

(ii) $(C_r(A))^+$ exists.

(iii) $Tr(C_r(A)^*C_r(A)) = \sum_{(\alpha,\beta)\in\Delta} \overline{|A_{\alpha\beta}|}|A_{\alpha\beta}|$ is a unit of $R$.

In case $\sum_{(\alpha,\beta)\in\Delta} \overline{|A_{\alpha\beta}|}|A_{\alpha\beta}| = u$ is a unit of $R$ then $G = (g_{ij})$ defined by $g_{ij} = u^{-1} \sum_{(\alpha,\beta)\in\Delta} \overline{|A_{\alpha\beta}|}\frac{\partial}{\partial a_{ji}}|A_{\alpha\beta}|$ is the Moore-Penrose inverse of $A$.

**Proof:** (i) $\Rightarrow$ (ii) follows from the properties of compound matrices given in section 2.5. In fact $C_r(A^+) = (C_r(A))^+$.

(ii) $\Rightarrow$ (iii). Since $\rho(C_r(A)) = 1$ from Theorem 6.6, it follows that $Tr(C_r(A^*)C_r(A)) = \sum_{(\alpha,\beta)\in\Delta} \overline{|A_{\alpha\beta}|}|A_{\alpha\beta}|$ is a unit of $R$.

(iii) $\Rightarrow$ (i). Let $u = Tr(C_r(A^*)C_r(A))$. Then, $\sum_{(\alpha,\beta)\in\Delta} u^{-1}\overline{|A_{\alpha\beta}|}|A_{\alpha\beta}| = 1$. We define $G = (g_{ij})$ by $g_{ij} = \sum_{(\alpha,\beta)\in\Delta} u^{-1}\overline{|A_{\alpha\beta}|}\frac{\partial|A_{\alpha\beta}|}{\partial a_{ji}}$ and show that $G$ is the Moore-Penrose inverse of $A$. We shall do this in an indirect way.

Let $F$ be the field of quotients of $R$ with $a \to \bar{a}$ extended to $F$. Consider $A$ as a matrix over $F$. From the hypothesis (iii) that $\sum_{(\alpha,\beta)\in\Delta} \overline{|A_{\alpha\beta}|}|A_{\alpha\beta}|$ is a unit of $R$ it follows that $\sum_{(\alpha,\beta)\in\Delta} \overline{|A_{\alpha\beta}|}|A_{\alpha\beta}|$ is also a unit of $F$. But over $F$, $A$ has a rank factorization. Hence Proposition 6.5 is applicable and we get that $A$ has Moore-Penrose inverse over $F$. Let it be $H$.

Then, $C_r(H)$ is the Moore-Penrose inverse of $C_r(A)$. On the other hand since $u = \sum_{(\alpha,\beta)\in\Delta} \overline{|A_{\alpha\beta}|}|A_{\alpha\beta}| = Tr(C_r(A)^*C_r(A))$ is a unit, by Theorem 6.6, $u^{-1}C_r(A^*)$ is the Moore-Penrose inverse of $C_r(A)$. By the uniqueness of the Moore-Penrose inverse (section 3.1) it follows that $C_r(H) = u^{-1}C_r(A^*)$. Thus $|H_{\beta\alpha}| = u^{-1}\overline{|A_{\alpha\beta}|}$. Also, by Theorem 6.3, since $H$ is a reflexive $g$-inverse of $A$,

$$\begin{aligned} h_{ij} &= \sum_{(\alpha,\beta)\in\Delta} |H_{\beta\alpha}|\frac{\partial|A_{\alpha\beta}|}{\partial a_{ji}} \\ &= \sum_{(\alpha,\beta)\in\Delta} u^{-1}\overline{|A_{\alpha\beta}|}\frac{\partial|A_{\alpha\beta}|}{\partial a_{ji}} = g_{ij}. \end{aligned}$$

Thus $H = G$. But $G$ is a matrix over $R$. Hence $H$ is a matrix over $R$ and it is also the Moore-Penrose inverse of $A$. Hence $G$ is the Moore-Penrose inverse of $A$. ♠

In the case of fields, a matrix $A$ has Moore-Penrose inverse if and only if $\rho(A^*A) = \rho(AA^*) = \rho(A)$. In the case of integral domains also a similar result holds.

**Theorem 6.8:** Let $A$ be an $m \times n$ matrix over an integral domain $R$ with involution $a \to \bar{a}$. Let $\rho(A) = r$. Then the following are equivalent.

(i) A has Moore-Penrose inverse.

(ii) $\sum_{(\alpha,\beta)\in\Delta} \overline{|A_{\alpha\beta}|}|A_{\alpha\beta}|$ is a unit of $R$.

(iii) $\rho(A^*A) = \rho(AA^*) = \rho(A)$ and $A^*A$ and $AA^*$ are regular.

**Proof:** (i) $\Rightarrow$ (ii) was already shown in the previous Theorem.

(ii) $\Rightarrow$ (iii). The matrix $C_r(A^*)C_r(A) = C_r(A^*A)$ is a nonzero matrix because $Tr(C_r(A^*)C_r(A)) = \sum_{(\alpha,\beta)\in\Delta} \overline{|A_{\alpha\beta}|}|A_{\alpha\beta}|$ is a unit of $R$. Hence $C_r(A^*A)$ is a nonzero matrix. This implies that $\rho(A^*A) \geq r$. However $\rho(A^*A) \leq \rho(A) = r$. Thus we get that $\rho(A^*A) = r = \rho(A)$. Since a linear combination of all the $r \times r$ minors of $A^*A$, namely,

$$\sum_{(\alpha,\beta)\in\Delta} \delta^{\alpha}_{\beta}|(A^*A)_{\alpha\beta}| = \sum_{\alpha}|(A^*A)_{\alpha\alpha}| = \sum_{\alpha}\sum_{\gamma}|A^*_{\alpha\gamma}||A_{\gamma\alpha}|$$

$$= \sum_{\alpha,\gamma}\overline{|A_{\gamma\alpha}|}|A_{\gamma\alpha}| = 1,$$

where $\delta^{\alpha}_{\beta} = 1$ if $\alpha = \beta$ and $= 0$ if $\alpha \neq \beta$, by Theorem 6.1, we have that $A^*A$ is regular. Similarly $AA^*$ is also regular.

(iii) $\Rightarrow$ (i). Proof of this implication is as in the case of fields. Let us verify that $A^*(AA^*)^-A(A^*A)^-A^*(= H$, say) is the Moore-Penrose inverse of $A$. Since $\rho(A^*A) = \rho(A) = \rho(AA^*)$, when $A$, $A^*A$ and $AA^*$ are considered as matrices over the field $F$ of quotients of $R$, we get matrices $B$ and $C$ over $F$ such that $BA^*A = A$ and $AA^*C = A$.

Hence $A(A^*A)^-A^*A = BA^*A(A^*A)^-A^*A = BA^*A = A$ and $AA^*(AA^*)^-A = A$. Now, $AHA = AA^*(AA^*)^-A = A$. Similarly one can also show that $HAH = H, (AH)^* = AH$ and$(HA)^* = HA$. Thus $A$ has Moore-Penrose inverse.

♠

## 6.4 Generalized inverses of the form $PCQ$

We have seen in the proof of Theorem 6.8 that the Moore-Penrose of a matrix $A$, when it exists, is of the form $A^*DA^*$ for some $D$ (In fact, $A^+ = A^*(AA^*)^-A(A^*A)^-A^*$). Taking the cue from here let us find necessary and sufficient conditions for a matrix $A$ to have a $g$-inverse of the form $PCQ$ for some matrix $C$, for given $P$ and $Q$. This turns out to be a very useful result.

**Proposition 6.9:** Let $A$ of order $m \times n$, $P$ of order $n \times k$, $Q$ of order $\ell \times m$ be matrices over an integral domain $R$.

(a) If $C$ is a $k \times \ell$ matrix over $R$, $PCQ$ is a $g$-inverse of $A$ if and only if (i) $\rho(QAP) = \rho(A)$ and (ii) $C$ is a $g$-inverse of $QAP$.

(b) For some $k \times \ell$ matrix $C$, $A$ has a $g$-inverse of the form $PCQ$ if and only if (i) $\rho(QAP) = \rho(A)$ and (ii) $QAP$ is regular.

In case $\rho(Q) = \rho(P) = \rho(A)$ and if $A$ has a $g$-inverse of the form $PCQ$ then the $g$-inverse of $A$ of the form $PCQ$ is unique.

**Proof:** (a) If $PCQ$ is a $g$-inverse of $A$ then $A = APCQA$. Hence $A = APCQAPCQA$. It follows that $\rho(A) \leq \rho(QAP)$. In any case, $\rho(QAP) \leq \rho(A)$. Thus $\rho(A) = \rho(QAP)$. Also, $A = APCQA$ gives us that $QAP = QAPCQAP$. Thus $C$ is a $g$-inverse of $QAP$.

To show the converse, let $F$ be the field of quotients of the integral domain $R$. Since $\rho(QAP) \leq \rho(QA) \leq \rho(A), \rho(QAP) \leq \rho(AP) \leq \rho(A)$ and $\rho(QAP) = \rho(A)$, we get that $\rho(QA) = \rho(A)$ and $\rho(AP) = \rho(A)$. Considering $QA, AP$ and $A$ as matrices over $F$, we get matrices $D$ and $E$ over $F$ such that $DQA = A$ and $APE = A$. Now, since $QAPCQAP = QAP$, we get, $DQAPCQAPE = DQAPE$. Thus $APCQA = A$ i.e., $PCQ$ is a $g$-inverse of $A$.

Now, let $\rho(Q) = \rho(P) = \rho(A)$. Since $A$ has a $g$-inverse of the form $PCQ$, by (a) above $\rho(QAP) = \rho(A)$. Hence $\rho(Q) = \rho(P) = \rho(QAP)$. This implies that there are matrices $D$ and $E$ over the field $F$ of quotients of $R$ such that $Q = QAPD$ and $P = EQAP$. Also, if $PCQ$ is a $g$-inverse of $A$ then by (a) above $C$ is a $g$-inverse of $QAP$. Hence, $PCQ = EQAPCQAPD = EQAPD = EQ$ and this is independent of $C$. Thus the matrix $PCQ$ is unique.

♠

Let us see some particular cases of the above Proposition.

**Proposition 6.10:** Let $R$ be an integral domain with an involution $a \to \bar{a}$. Let $A$ be an $m \times n$ matrix over $R$.

(a) If $C$ is an $n \times n$ matrix then $CA^*$ is a $g$-inverse of $A$ if and only if (i) $\rho(A^*A) = \rho(A)$ and (ii) $C$ is a $g$-inverse of $A^*A$.

(b) For some $n \times n$ matrix $C$, $A$ has a $g$-inverse of the form $CA^*$ if and only if (i) $\rho(A^*A) = \rho(A)$ and (ii) $A^*A$ is regular.

The corresponding result for matrices of the form $A^*D$ for some $m \times m$ matrix $D$ also holds.

**Proof:** In Proposition 6.9 if we take $P = I$ and $Q = A^*$ we get the result.

♠

**Proposition 6.11:** Let $R$ be an integral domain with an involution $a \to \bar{a}$. Let $A$ be a $m \times n$ matrix over $R$.

(a) If $C$ is an $m \times n$ matrix then $A^*CA^*$ is a $g$-inverse of $A$ if and only if (i) $\rho(A^*AA^*) = \rho(A)$ and (ii) $C$ is a $g$-inverse of $A^*AA^*$.

(b) For some $m \times n$ matrix $C$, $A$ has a $g$-inverse of the form $A^*CA^*$ if and only if (i) $\rho(A^*AA^*) = \rho(A)$ and (ii) $A^*AA^*$ is regular.

**Proof:** In Proposition 6.9 if we take $P = A^*$ and $Q = A^*$ we get the result.

♠

**Proposition 6.12:** Let $A$ be an $m \times m$ matrix over an integral domain $R$.

(a) If $C$ is an $m \times m$ matrix then $AC$ is a $g$-inverse of $A$ if and only if (i) $\rho(A^2) = \rho(A)$ and (ii) $C$ is a $g$-inverse of $A^2$.

(b) For some $m \times m$ matrix $C$, $A$ has a $g$-inverse of the form $AC$ if and only if (i) $\rho(A^2) = \rho(A)$ and (ii) $A^2$ is regular.

The corresponding result for matrices of the form $CA$ also holds. Consequently, $AC$ is a $g$-inverse of $A$ if and only if $CA$ is a $g$-inverse of $A$.

**Proof:** In Proposition 6.9 if we take $P = A$ and $Q = I$ we get the result.

♠

For the existence of $g$-inverses of $A$ of the form $ACA$ we have

**Proposition 6.13:** Let $A$ be an $m \times m$ matrix over an integral domain $R$.

(a) If $C$ is an $m \times m$ matrix, then $ACA$ is a $g$-inverse of $A$ if and only if (i) $\rho(A^3) = \rho(A)$ and (ii) $C$ is a $g$-inverse of $A^3$.

(b) For some $m \times m$ matrix $C$, $A$ has a $g$-inverse of the form $ACA$ if and only if (i) $\rho(A^3) = \rho(A)$ and (ii) $A^3$ is regular.

Further, if $A$ has a $g$-inverse of the form $ACA$ then $A$ has a unique $g$-inverse of the form $ACA$.

**Proof:** In Proposition 6.9 if we take $P = A$ and $Q = A$ we get the result.

♠

We have already seen in Proposition 6.12 that $CA$ is a $g$-inverse of $A$ if $AC$ is a $g$-inverse of $A$. It can also be seen that if $AD$ and $EA$ are $g$-inverses of $A$ then $ADAEA$ is also a $g$-inverse of $A$ and it is of the form $ACA$. We shall see now that in fact Proposition 6.12 and Proposition 6.13 give the same result.

**Proposition 6.14:** Let $A$ be an $m \times m$ matrix over an integral domain $R$. Let $n \geq 2$. The following are equivalent.

(i) $A$ has a $g$-inverse of the form $AC$.

(ii) $A$ has a $g$-inverse of the form $CA$.

(iii) $\rho(A) = \rho(A^2)$ and $A^2$ is regular.
(iv) $A$ has a $g$-inverse of the form $ACA$.
(v) $\rho(A) = \rho(A^3)$ and $A^3$ is regular.
(vi) $\rho(A) = \rho(A^n)$ and $A^n$ is regular.

**Proof:** (i) $\Leftrightarrow$ (ii) $\Leftrightarrow$ (iii) is Proposition 6.12. (iv) $\Leftrightarrow$ (v) is Proposition 6.13.

(v) or (vi) $\Rightarrow$ (i). Let $G$ be a $g$-inverse of $A^n$. Since $\rho(A) = \rho(A^n)$ there is a matrix $B$ over the field $F$ of quotients of $R$ such that $A = A^nB$. Since $A^nGA^n = A^n$ we get that $A^nGA = A$. Hence $A(A^{n-1}G)A = A$. Since $n \geq 2$ we get that $A$ has a $g$-inverse of the form $AC(= AA^{n-2}G)$.

(iii) $\Rightarrow$ (v) or (vi). Since $\rho(A) = \rho(A^2)$ and since $R$ is an integral domain there is a matrix $B$ over the field $F$ of quotients of $R$ such that $A = A^2B$. Let $G$ be a $g$-inverse of $A^2$, i.e., $A^2GA^2 = A^2$. Hence $A^2GA = A$. Now, $A^n(GA)^{n-1} = A^{n-2}A^2GA(GA)^{n-2} = A^{n-1}(GA)^{n-2} = \cdots = A^2GA = A$. Hence $A^n = AA^{n-1} = A^n(GA)^{n-1}A^{n-1} = A^n(GA)^{n-2}GAA^{n-1} = A^n(GA)^{n-2}GA^n$. Thus $(GA)^{n-2}G$ is a $g$-inverse of $A^n$. Also, for the $B$ as above we have $A = A^2B = A^3B^2 = \cdots = A^nB^{n-1}$. Hence $\rho(A) \leq \rho(A^n)$, but $\rho(A^n) \leq \rho(A)$ holds always. Thus $\rho(A) = \rho(A^n)$.

♠

This completes our study of $g$-inverses of the form $PCQ$.

## 6.5 $\{1,2,3\}$- and $\{1,2,4\}$-inverses

We shall now look at conditions for the existence of $\{1,2,3\}$ (and $\{1,2,4\}$)-inverses of matrices over integral domains. We shall also give a construction of a $\{1,2,3\}$-inverse when it exists and describe all $\{1,2,3\}$-inverses.

Recall from section 3.1 that a matrix $G$ is called a $\{1,2,3\}$-inverse of a matrix $A$ if

$$AGA = A \tag{1}$$
$$GAG = G \tag{2}$$
$$\text{and} \quad (AG)^* = AG \tag{3}$$

are satisfied. Similarly, a matrix $G$ is called a $\{1,2,4\}$-inverse of a matrix $A$ if in addition to (1) and (2) $A$ and $G$ satisfy $(GA)^* = GA$. We shall present a comprehensive result that gives necessary and sufficient conditions for the existence of $\{1,2,3\}$-inverses.

First we shall characterize all $\{1, 2, 3\}$-inverses (and $A^+$) of a given matrix $A$ among all its $g$-inverses.

**Proposition 6.15:** Let $A$ be an $m \times n$ matrix over an integral domain $R$ with an involution $a \to \bar{a}$. Then

(a) A $g$-inverse $G$ of $A$ is a $\{1, 2, 3\}$-inverse of $A$ if and only if $G$ is of the form $CA^*$ for some $C$.

(b) A $g$-inverse $G$ of $A$ is a $\{1, 2, 4\}$-inverse of $A$ if and only if $G$ is of the form $A^*C$ for some $C$.

(c) A $g$-inverse $G$ of $A$ is the Moore-Penrose inverse of $A$ if and only if $G$ is of the form $A^*CA^*$ for some $C$.

**Proof:** (a) If $G$ is a $\{1, 2, 3\}$-inverse of $A$ then $AGA = A, GAG = G$ and $(AG)^* = G^*A^* = AG$. So, $GG^*A^* = G$. Hence $G$ is a $g$-inverse of the form $CA^*$.

Conversely, if $G$ is a $g$-inverse of $A$ and also $G = CA^*$ for some $C$ then $ACA^*A = A$. Hence $ACA^*AC^*A^* = AC^*A^*$, i.e., $(ACA^*)(ACA^*)^* = (ACA^*)^*$. Hence $(ACA^*)^* = ACA^*$. Thus $(AG)^* = AG$. Now, $GAG = CA^*ACA^* = C(ACA^*A)^* = CA^*G$. Thus $G$ is a $\{1, 2, 3\}$-inverse of $A$. Parts (b) and (c) are proved along the same lines.

♠

The comprehensive result is the following.

**Theorem 6.16:** Let $A$ be an $m \times n$ matrix over an integral domain $R$ with an involution $a \to \bar{a}$. Let $\rho(A) = r$. Then the following are equivalent.

(i) $A$ has $\{1, 3\}$-inverse.

(ii) $C_r(A)$ has $\{1, 3\}$-inverse.

(iii) $C_r(A)$ has $\{1, 2, 3\}$-inverse.

(iv) There is a matrix $T$ such that $T^* = T$ and $Tr(C_r(A)TC_r(A^*)) = 1$.

(v) A linear combination of all the $r \times r$ minors of $A^*A$ is equal to one.

(vi) $\rho(A^*A) = \rho(A)$ and $A^*A$ is regular.

**Proof:** (i) $\Rightarrow$ (ii) follows from the properties of compound matrices given in section 2.5. In fact, if $G$ is a $\{1, 3\}$-inverse of $A, C_r(G)$ is a $\{1, 3\}$-inverse of $C_r(A)$.

(ii) $\Rightarrow$ (iii) follows from the observations in section 3.1. In fact, if $S$ is a $\{1, 3\}$-inverse of $C_r(A)$ then $SC_r(A)S$ is a $\{1, 2, 3\}$-inverse of $C_r(A)$.

(iii) $\Rightarrow$ (iv). Let $S$ be a $\{1, 2, 3\}$-inverse of $C_r(A)$. Since $\rho(C_r(A)) = 1$, by Theorem 5.2 we have that $Tr(C_r(A)S) = 1$. Now, let $T = SS^*$. Then, $C_r(A)TC_r(A^*) = C_r(A)SS^*C_r(A^*) = C_r(A)S(C_r(A)S)^* = C_r(A)C_rS(A)S = C_r(A)S$. Hence $Tr(C_r(A)TC_r(A^*)) = Tr(C_r(A)S) = 1$ and $T^* = T$.

(iv) $\Rightarrow$ (v). Since $Tr(C_r(A)TC_r(A^*)) = 1$, we have that $Tr(C_r(A^*A)T) = Tr(C_r(A^*)C_r(A)T) = Tr(C_r(A)TC_r(A^*)) = 1$. This means that a linear combination of all the $r \times r$ minors of $A^*A$ is equal to one.

(v) $\Rightarrow$ (vi). Since some linear combination of all the $r \times r$ minors of $A^*A$ is equal to one, at least one $r \times r$ minor of $A^*A$ is nonzero. This implies that $\rho(A^*A) \geq r$. But $\rho(A^*A) \leq \rho(A) = r$. Thus $\rho(A^*A) = r$. Now, Theorem 6.1 implies that $A^*A$ is regular.

(vi) $\Rightarrow$ (i). Since $\rho(A^*A) = \rho(A)$ and $A^*A$ is regular, by Theorem 6.10, if $C$ is a $g$-inverse of $A^*A$, $CA^*$ is a $g$-inverse of $A$. By Proposition 6.15, $CA^*$ is a $\{1,2,3\}$-inverse of $A$.

♠

We shall now give methods of constructing $\{1,2,3\}$-inverses when they exist.

**Proposition 6.17:** Let $A$ be an $m \times n$ matrix over an integral domain $R$ with an involution $a \to \bar{a}$. Let $\rho(A) = r$.

(a) If $S = (s_{\beta\alpha})_{(\alpha,\beta)\in\Delta}$ is an $\binom{n}{r} \times \binom{m}{r}$ matrix which is a $\{1,3\}$-inverse of $C_r(A)$ then $Tr(C_r(A)S) = \sum_{(\alpha,\beta)\in\Delta} |A_{\alpha\beta}|s_{\beta\alpha} = 1$ and $G = (g_{ij})$ defined by $g_{ij} = \sum_{(\alpha,\beta)\in\Delta} \frac{\partial|A_{\alpha\beta}|}{\partial a_{ji}} s_{\beta\alpha}$ is a $\{1,3\}$-inverse of $A$.

(b) If $S = (s_{\beta\alpha})_{(\alpha,\beta)\in\Delta}$ is an $\binom{n}{r} \times \binom{m}{r}$ matrix which is a $\{1,2,3\}$-inverse of $C_r(A)$ then $Tr(C_r(A)S) = \sum_{(\alpha,\beta)\in\Delta} |A_{\alpha\beta}|s_{\beta\alpha} = 1$ and $G = (g_{ji})$ defined by $g_{ij} = \sum_{(\alpha,\beta)\in\Delta} \frac{\partial|A_{\alpha\beta}|}{\partial a_{ji}} s_{\beta\alpha}$ is a $\{1,2,3\}$-inverse of $A$.

(c) If $T = (t_{\beta\alpha})_{(\alpha,\beta)\in\Delta}$ is an $\binom{n}{r} \times \binom{m}{r}$ matrix such that $Tr(C_r(A)TC_r(A^*)) = \sum_{\substack{\beta\in Q_{m,r}\\ \gamma\in Q_{n,r}\\ \alpha\in Q_{n,r}}} |A_{\beta\gamma}|t_{\gamma\alpha}|\bar{A}_{\beta\alpha}| = 1$ then $G = (g_{ij})$ defined by $g_{ij} = \sum_{\substack{\beta\in Q_{m,r}\\ \gamma\in Q_{n,r}\\ \alpha\in Q_{n,r}}} \frac{\partial|A_{\beta\gamma}|}{\partial a_{ji}} t_{\gamma\alpha}|\bar{A}_{\beta\alpha}|$ is a $\{1,2,3\}$-inverse of $A$.

(d) If $\rho(A^*A) = \rho(A)$ and $A^*A$ is regular then $G = (A^*A)^-A^*$ where $(A^*A)^-$ is any $g$-inverse of $A^*A$, is a $\{1,2,3\}$-inverse of $A$.

**Proof:** (a) Let $S = (s_{\beta\alpha})_{(\alpha,\beta)\in\Delta}$ be an $\binom{n}{r} \times \binom{m}{r}$ matrix which is a $\{1,3\}$-inverse of $C_r(A)$. Since $\rho(C_r(A)) = 1$ by Theorem 5.2 we have that $Tr(C_r(A)S) = \sum_{(\alpha,\beta)\in\Delta} |A_{\alpha\beta}|s_{\beta\alpha} = 1$. Also, if $G = (g_{ij})$ is defined by $g_{ij} = \sum_{(\alpha,\beta)\in\Delta} \frac{\partial|A_{\alpha\beta}|}{\partial a_{ji}} s_{\beta\alpha}$, then $G$ is a $g$-inverse of $A$ by Theorem 5.4. $(AG)^* = AG$ follows because of $(C_r(A)S)^* = C_r(A)S$ from Theorem 5.5.

(b) Let $S = (s_{\beta\alpha})_{(\alpha,\beta)\in\Delta}$ be an $\binom{n}{r} \times \binom{m}{r}$ matrix which is a $\{1,2,3\}$-inverse of $C_r(A)$. Let $G = (g_{ij})$ be defined as in the proof of (a). Since $C_r(A)SC_r(A) = C_r(A)$ and $r \geq 1$ we have that $S$ is a nonzero matrix. Since $SC_r(A)S = S$ we have that $\rho(S) \leq \rho(C_r(A)) = 1$. Hence $\rho(S) = 1$.

By Theorem 6.3, we get that $G$ is a reflexive $g$-inverse of $A$. By (a) above we get that $G$ is $\{1,3\}$-inverse of $A$. Hence $G$ is a $\{1,2,3\}$-inverse of $A$.

(c) If we consider $TC_r(A^*)$ as the matrix $S$ of (a) above, then, since $Tr(C_r(A)TC_r(A^*)) = 1$, we get that $S \neq 0$. Also, $\rho(S) \leq \rho(C_r(A^*)) = 1$. Hence $\rho(S) = 1$. Also $S$ is a $\{1,3\}$-inverse of $C_r(A)$. That $S$ is a $g$-inverse of $C_r(A)$ follows because $Tr(C_r(A)S) = 1$ (by Theorem 5.1 (b)). That, $(C_r(A)S)^* = (C_r(A)TC_r(A^*))^* = (C_r(A)T^*C_r(A^*)) = C_r(A)TC_r(A^*)$ follows because $T = T^*$. Thus $S$ is a $\{1,2,3\}$-inverse of $C_r(A)$. Hence $G$ defined from this $S$ is a $\{1,2,3\}$-inverse of $A$.

(d) The proof of this was already included in the proof of (vi) $\Rightarrow$ (i) of Theorem 6.16.

♠

In case a matrix $A$ has a rank factorization $LR$ then the existence of $\{1,2,3\}$-inverse of $A$ can be characterized in terms of $L$.

**Proposition 6.18:** Let $A$ be an $m \times n$ matrix over an integral domain $R$ with an involution $a \to \bar{a}$. Let $\rho(A) = r$.

(a) If $A = LD$ is a rank factorization of $A$ then $A$ has $\{1,2,3\}$-inverse if and only if $L^*L$ is unimodular.

(b) If $A = LD$ is a factorization such that $L$ is $m \times r$ and $D$ is $r \times n$ then $A$ has $\{1,2,3\}$-inverse if and only if $L^*L$ is unimodular and $D$ has a right inverse over the integral domain $R$.

**Proof:** We shall prove (b). (a) follows from (b) easily.

Let $F$ be the field of quotients of $R$.

Let $G$ be a $\{1,2,3\}$-inverse of $A$. Hence $LDGLD = LD$. Since $L$ has a left inverse over $F$ and $D$ has a right inverse over $F$ we get that $DGL = I$. Thus $D$ has a right inverse over $R$. Also $(AG)^* = (LDG)^* = LDG$, i.e., $G^*D^*L^* = LDG$. Hence $G^*D^*L^*L = LDGL = L$. Hence $DGG^*D^*L^*L = DGL = I$. Thus $L^*L$ is unimodular.

If $L^*L$ is unimodular and $D$ has a right inverse, say $D_R^{-1}$, then, let us see that $G = D_R^{-1}(L^*L)^{-1}L^*$ is a $\{1,2,3\}$-inverse of $A = LD$.

$AGA = LDGLD = LDD_R^{-1}(L^*L)^{-1}L^*LD = LD = A$. $GAG = GLDG = D_R^{-1}(L^*L)^{-1}L^*LDD_R^{-1}(L^*L)^{-1}L^* = D_R^{-1}(L^*L)^{-1}L^* = G$.

$(AG)^* = (LDD_R^{-1}(L^*L)^{-1}L^*)^* = LDG = AG$.

♠

Similar results also hold for $\{1,2,4\}$-inverses and Moore-Penrose inverses. For example, it follows that an $m \times n$ matrix $A$ over an integral domain $R$ with involution $a \to \bar{a}$ will have Moore-Penrose inverse if and only if $\rho(A^*A) = \rho(A) = \rho(AA^*)$, and $A^*A$ and $AA^*$ are regular.

## 6.6 Group inverses over integral domains

Let $A$ be an $m \times m$ matrix over a commutative ring. Recall that an $m \times m$ matrix $G$ is called a group inverse of $A$ if

$$AGA = A \tag{1}$$

$$GAG = G \tag{2}$$

$$\text{and } AG = GA \tag{5}$$

are satisfied. If $G$ satisfies only (1) and (5) then $G$ is called a commuting $g$-inverse of $A$. Clearly, if $G$ is a commuting $g$-inverse of $A$, $GAG$ is a group inverse of $A$. Thus a square matrix $A$ has a group inverse if and only if it has a commuting $g$-inverse. Also, the group inverse when it exists is unique. In fact, if $G$ and $H$ both are group inverses of $A$, then $G = GAG = GAHAG = AGAHG = AHG$ and also $H = HAH = HAGAH = HGAHA = HGA = HAG = AHG$. Hence $G = H$.

Also, observe that if $G$ is a group inverse of $A$ then, $G = GAG = AGG$. Thus $A$ has a $g$-inverse of the form $AC$ for some square matrix $C$. This observation is the basis of a characterization of matrices which admit group inverses for matrices over integral domains.

**Theorem 6.19:** Let $A$ be a square matrix over an integral domain $R$. Let $n$ be an integer $\geq 2$. The following are equivalent.

(i) $A$ has a group inverse.
(ii) $A$ has a $g$-inverse of the form $AC$ for some $C$.
(iii) $A$ has a $g$-inverse of the form $CA$ for some $C$.
(iv) $A$ has a $g$-inverse of the form $ACA$ for some $C$.
(v) $\rho(A) = \rho(A^2)$ and $A^2$ is regular.
(vi) $\rho(A) = \rho(A^n)$ and $A^n$ is regular.

Further, if $ACA$ is a $g$-inverse of $A$ then $ACA$ is the group inverse of $A$.

**Proof:** We have already observed in Proposition 6.14 that all of (ii)$-$(vi) are equivalent. (i) $\Rightarrow$ (ii) was observed just preceding the statement of Theorem 6.19. We shall now show (iv) $\Rightarrow$ (i). We shall in fact show that if $ACA$ is a $g$-inverse of $A$ then $ACA$ is a group inverse of $A$.

Let $G = ACA$ we have $A^2CA^2 = AGA = A$. Hence $A^2CAA = A$. Hence $\rho(A) = \rho(A^2)$ and this gives us a matrix $D$ over the field $F$ of quotients of $R$ such that $DA^2 = A$. From $A^2CAA = A$ we get $A^2CAA^2 = A^2$. Using the matrix $D$ we get $ACAA^2 = A$.

Also note that $A^2CA^2 = A$ can be rewritten as $AACA^2 = A$. Hence $A^2ACA^2 = A^2$. Since $\rho(A) = \rho(A^2)$, we get a matrix $E$ over the field $F$ of quotients of $R$ such that $A^2E = A$. We get with the help of $E$ that $A^2ACA = A$.

Thus we have shown that $ACAA^2 = A$ and $A^2ACA = A$ hold.

Now, $GAG = ACAAACA = ACA = G$, and $AG = AACA = ACAA^2ACA = ACAA = GA$.

Thus $G$ is the group inverse of $A$.

♠

In the above proof we have shown that a $g$-inverse of $A$ of the form $ACA$ is necessarily the group inverse of $A$. This amounts to saying that if $\rho(A) = \rho(A^3)$ and $C$ is a $g$-inverse of $A^3$ then $ACA$ is the group inverse of $A$. However, in general, a $g$-inverse of $A$ of the form $AC$ (or $CA$) need not be a group inverse of $A$. Let $A = \begin{bmatrix} 1 & 2 \\ 1 & 2 \end{bmatrix}$ be a matrix over the field of real numbers. If we take $C = \begin{bmatrix} 1/3 & 0 \\ 0 & 0 \end{bmatrix}$, $AC = \begin{bmatrix} 1/3 & 0 \\ 1/3 & 0 \end{bmatrix}$ and $AC$ is a $g$-inverse of $A$ but $AC$ is not the group inverse of $A$.

If $A$ has group inverse and $\rho(A) = r$ from the properties of compound matrices it can easily be seen that $C_r(A)$ has a group inverse. The converse also holds. We shall prove this and at the same time give methods of construction of the group inverse when it exists in the next Theorem.

**Theorem 6.20:** Let $A$ be an $m \times m$ matrix over an integral domain $R$ with $\rho(A) = r$. Let $u = \sum_{\gamma \in Q_{r,m}} |A_{\gamma\gamma}| = Tr(C_r(A))$. Then the following are equivalent

(i) $A$ has group inverse.

(ii) $C_r(A)$ has group inverse.

(iii) $u$ is a unit.

(iv) $u^{-2} \sum_{\alpha \in Q_{r,m}, \beta \in Q_{r,m}} |A_{\alpha\beta}||A_{\beta\alpha}| \left( = u^{-2} \sum_{\alpha \in Q_{r,m}} |(A^2)_{\alpha\alpha}| \right) = 1.$

In case $u$ is a unit $G = (g_{ij})$ defined by

$$g_{ij} = \sum_{\gamma \in Q_{r,m}} \frac{\partial |A_{\gamma\gamma}|}{\partial a_{ji}}$$

is a commuting $g$-inverse of $A$.

In case $u^{-2} \sum_{\alpha \in Q_{r,m}, \beta \in Q_{r,m}} |A_{\alpha\beta}||A_{\beta\alpha}| = 1, G = (g_{ij})$ defined by

$$g_{ij} = \sum_{\alpha \in Q_{r,m}, \beta \in Q_{r,m}} u^{-2} |A_{\beta\alpha}| \frac{\partial |A_{\alpha\beta}|}{\partial a_{ji}}$$

is the group inverse of $A$.

**Proof:** (i) $\Rightarrow$ (ii) is clear.

(ii) $\Rightarrow$ (iii). From Theorem 6.19 it follows that, since $C_r(A)$ has group inverse, $\rho(C_r(A^2)) = 1$ and $C_r(A^2)$ is regular. From Theorem 5.2 there is an $\binom{m}{r} \times \binom{m}{r}$ matrix $C = (c_{\beta\alpha})_{\beta\in Q_{r,m},\alpha\in Q_{r,m}}$ such that $Tr(C_r(A^2)C) = \sum_{\alpha\in Q_{r,m},\beta\in Q_{r,m}} |(A^2)_{\alpha\beta}|c_{\beta\alpha} = 1$. But by the Cauchy-Binet Theorem we have $|(A^2)_{\alpha\beta}| = \sum_{\gamma\in Q_{r,m}} |A_{\alpha\gamma}||A_{\gamma\beta}|$. Also, since $\rho(C_r(A)) = 1$ we have $|A_{\alpha\gamma}|A_{\gamma\beta}| = |A_{\gamma\gamma}||A_{\alpha\beta}|$ for all $\alpha, \beta$ and $\gamma$. Thus,

$$\sum_{\alpha\in Q_{r,m},\beta\in Q_{r,m},\gamma\in Q_{r,m}} |A_{\gamma\gamma}||A_{\alpha\beta}|c_{\beta\alpha}$$
$$= \left(\sum_{\gamma\in Q_{r,m}} |A_{\gamma\gamma}|\right)\left(\sum_{\alpha\in Q_{r,m}\beta\in Q_{r,n}} |A_{\alpha\beta}|c_{\beta\alpha}\right) = 1.$$

This shows that $u = \sum_{\gamma\in Q_{r,m}} |A_{\gamma\gamma}|$ is a unit.

If $u$ defined above is a unit, we have that $\sum_{\gamma\in Q_{r,m}} u^{-1}|A_{\gamma\gamma}| = 1$. But $\sum_{\gamma\in Q_{r,m}} u^{-1}|A_{\gamma\gamma}| = Tr(C_r(A)u^{-1}I)$ where $I$ is the $\binom{m}{r} \times \binom{m}{r}$ identity matrix. Hence $Tr(C_r(A)u^{-1}I) = 1$. Using the notation of section 5.3, if we write $G = A_{u^{-1}I}$ then $AGA = A$ because $Tr(C_r(A)u^{-1}I) = 1$. Also, $AG = GA$ because of Theorem 5.5 (e) and because $C_r(A)u^{-1}I = u^{-1}IC_r(A)$. But $A_{u^{-1}I}$ is the matrix $(g_{ij})$ where $g_{ij} = \sum_{\gamma\in Q_{r,m}} u^{-1}\frac{\partial|A_{\gamma\gamma}|}{\partial a_{ji}}$. Thus $G$ is a commuting $g$-inverse of $A$.

(iii) $\Rightarrow$ (iv). $u$ being a unit, $\sum_{\gamma\in Q_{r,m}} u^{-1}|A_{\gamma\gamma}| = Tr(C_r(A)u^{-1}I) = 1$. Since $\rho(C_r(A)) = 1$ by Theorem 5.2 we get that $C_r(A)u^{-1}IC_r(A) = C_r(A)$. If we multiply with $u^{-1}$ we obtain $C_r(A)u^{-2}C_r(A) = u^{-1}C_r(A)$. Thus we have $Tr(u^{-2}C_r(A)C_r(A)) = 1$. But $Tr(u^{-2}C_r(A)C_r(A)) = \sum_{\alpha\in Q_{r,m},\beta\in Q_{r,m}} u^{-2}|A_{\alpha\beta}||A_{\beta\alpha}|$. Thus (iv) is proved.

Now, if $Tr(C_r(A)u^{-2}C_r(A)) = 1$ then, in the notation of section 5.3, $G = A_{u^{-2}C_r(A)}$ satisfies $AGA = A$ because of Theorem 5.5 (a), satisfies $GAG = G$ because of Theorem 5.5 (b) and the fact that $\rho(u^{-2}C_r(A)) = 1$ and satisfies $AG = GA$ because of Theorem 5.5 (e) and the fact that $C_r(A)u^{-2}C_r(A) =$

$u^{-2}C_r(A)C_r(A)$. Thus $A_{u^{-2}C_r(A)} = u^{-2}A_{C_r(A)}$ is the group inverse of $A$. But, if $u^{-2}A_{C_r(A)} = (g_{ij})$, by definition of $A_S$, $g_{ij} = \sum_{\alpha \in Q_{r,m}, \beta \in Q_{r,m}} u^{-2}|A_{\beta\alpha}|\frac{\partial |A_{\alpha\beta}|}{\partial a_{ji}}$.

♠

The existence of Moore-Penrose inverse of a matrix $A$ can be related to the existence of group inverse of $A^*A$. This is given as an Exercise at the end of the Chapter.

In the proof of the previous Theorem we have shown that if $A$ has a group inverse then $u = \sum_{\gamma \in Q_{r,m}} |A_{\gamma\gamma}|$ is a unit, $A_{u^{-1}I}$ is a commuting $g$-inverse of $A$ and that $A_{u^{-2}C_r(A)}$ is the group inverse of $A$.

Over an integral domain also, as in the case of fields, if a matrix $A$ has a group inverse then the group inverse is a polynomial in $A$ with coefficients from the integral domain. We shall show this now.

**Theorem 6.21:** If $A$ is an $m \times m$ matrix over an integral domain $R$ such that the group inverse of $A$ exists then the group inverse of $A$ is a polynomial in $A$ with coefficients from $R$. More precisely, if $A$ is an $m \times m$ matrix over a commutative ring such that the group inverse of $A$ exists and if $Tr(C_r(A)) = \sum_{\gamma \in Q_{r,m}} |A_{\gamma\gamma}|$ is a unit then the group inverse of $A$ is a polynomial in $A$ with coefficients from $R$.

**Proof:** We shall in fact prove the second part of the Theorem. Consider the characteristic polynomial $|\lambda I - A|$ of $A$.

Let $|\lambda I - A| = p_r\lambda^{n-r} + p_{r-1}\lambda^{n-r+1} + \cdots + p_1\lambda^{n-1} + \lambda^n$ where $p_k = (-1)^k \sum_{\gamma \in Q_{k,m}} |A_{\gamma\gamma}|$ for $1 \leq k \leq r$. Since Cayley-Hamilton Theorem holds for matrices over integral domains (in fact, Cayley-Hamilton Theorem holds for matrices over any commutative ring, see [25]) we have

$$p_r A^{n-r} + p_{r-1}A^{n-r+1} + \cdots + p_1 A^{n-1} + A^n = 0.$$

Since $A$ has a group inverse, $p_r = \sum_{\gamma \in Q_{r,m}} |A_{\gamma\gamma}|$ is a unit. Hence we can write

$$A^{n-r} = q_{r-1}A^{n-r+1} + q_{r-2}A^{n-r+2} + \cdots + q_1 A^{n-1} + q_0 A^n$$

where for $1 \leq k \leq r, q_k = -p_k p_r^{-1}$ and $q_0 = -p_r^{-1}$. If $G$ is the group inverse of $A$ we have $A^{k+1}G^k = A$, $A^kG^{k+1} = G$ and $A^kG^k = AG$ for all $1 \leq k$. Hence multiplying the above equation with $G^{n-r+1}$ we get

$$G = q_{r-1}AG + q_{r-2}A + q_{r-3}A^2 + \cdots + q_1 A^{r-2} + q_0 A^{r-1}$$

From this, by multiplying with $A$, we get

$$AG = q_{r-1}A + q_{r-2}A^2 + q_{r-3}A^3 + \cdots + q_1 A^{r-1} + q_0 A^r.$$

Substituting this back into the equation for $G$ we get

$$G = (q_{r-1}^2 + q_{r-2})A + (q_{r-1}q_{r-2} + q_{r-3})A^2 + \cdots + (q_{r-1}q_1 + q_0)A^{r-1} + q_{r-1}q_0 A^r.$$

This is a polynomial in $A$ with coefficients from the integral domain. Thus the Theorem is proved.

♠

## 6.7 Drazin inverses over integral domains

For an $m \times m$ matrix $A$ over a commutative ring $R$, the defining equations for the group inverse $G$ of $A$ can be written as

$$A^2G = A \quad (1)$$
$$GAG = G \quad (2)$$
$$\text{and } AG = GA \quad (5)$$

Generalizing these equations, recall from section 3.1, that a matrix $G$ is called a Drazin inverse of $A$ if for some $k$

$$A^{k+1}G = A^k \quad (1^k)$$
$$GAG = G \quad (2)$$
$$\text{and } AG = GA \quad (5)$$

are satisfied.

If $G$ and $A$ satisfy $(1^k)$, $\rho(A^{k+1})$ must equal $\rho(A^k)$. This implies that the *index* $p$ of $A$ (defined as the smallest integer $p \geq 1$ such that $\rho(A^{p+1}) = \rho(A^p)$) is smaller than or equal to $k$. In fact,

**Proposition 6.22:** Over an integral domain $R$, if $A$ and $G$ satisfy $(1^k), (2)$ and $(5)$ then $A$ and $G$ satisfy $(1^p), (2)$ and $(5)$ where $p$ is the index of $A$. Also, if $A$ and $G$ satisfy $(1^p), (2)$ and $(5)$ where $p$ is the index of $A$, then $A$ and $G$ satisfy $(1^k), (2)$ and $(5)$ for all $k \geq p$.

**Proof:** Let $F$ be the field of quotients of $R$. If $A$ and $G$ satisfy $(1^k)$ then from the definition of the index we have that $p \leq k$ and $\rho(A^p) = \rho(A^k)$. Hence there is a matrix $D$ over $F$ such that $A^p = DA^k$. This gives us that $A^{p+1}G = A^p$. The second part is trivial.

♠

Thus, if $A$ has a Drazin inverse $G$ then $A$ and $G$ satisfy $(1^p), (2)$ and $(5)$. If $A$ and $G$ satisfy $(1^p), (2)$ and $(5)$ then we have $A^{p+1}G = A^p, G^2A = G$ and $AG = GA$. This implies that $A^{p+k}G^k = A^p$; $G^{k+1}A^k = G, (GA)^k = G^kA^k = GA = AG = A^kG^k = (AG)^k$ for all $k \geq 1$ and any power of $A$ commutes with any power of $G$.

Let us now see the uniqueness of the Drazin inverse when it exists.

**Proposition 6.23:** Over an integral domain (or, even over a commutative ring) $R$, if an $m \times m$ matrix $A$ has a Drazin inverse then it is unique.

**Proof:** Let $G$ and $H$ be Drazin inverses of $A$. Hence both the pairs $G, A$ and $H, A$ satisfy $(1^p), (2)$ and $(5)$. Thus we have

$$\begin{aligned} G = G^{p+1}A^p &= G^{p+1}A^{2p+1}H^{p+1} \\ &= A^pH^{p+1} = H. \end{aligned}$$

Thus the Drazin inverse when it exists is unique.

♠

If $G$ is the Drazin inverse of $A$ then $A^{p+1}G = A^p$. This implies that $A^{p+1}GA = A^{p+1}$. But $GA = G^{p+1}A^{p+1}$. Thus $A^{p+1}$ is regular if $A$ has the Drazin inverse. The converse is also true. In fact more is true.

**Proposition 6.24:** Let $A$ be an $m \times m$ matrix over an integral domain $R$. Let $p$ be the index of $A$. Then,

(a) If $A$ has Drazin inverse then $A^{p+1}$ is regular.

(b) For a $k \geq 1$, if $A^{p+k}$ is regular, then $A^{p+k+1}$ is regular. Also $A^{p+k-1}$ is regular.

(c) If $A^{2p+1}$ is regular then $A$ has Drazin inverse.

(d) If for a $k \geq 1$, $A^{p+k}$ is regular, then $A$ has Drazin inverse.

Further, if $A^{2p+1}$ is regular then $A^p(A^{2p+1})^-A^p$ is the Drazin inverse of $A$.

**Proof:** (a) was already proved before the statement of Proposition 6.24.

To prove (b) let $G$ be a $g$-inverse of $A^{p+k}$, i.e., $A^{p+k}GA^{p+k} = A^{p+k}$. Since $k \geq 1$ and $\rho(A^{p+k}) = \rho(A^{p+k-1})$ there is a matrix $E$ over the field $F$ of quotients of $R$ such that $A^{p+k}E = A^{p+k-1}$. Hence we have $A^{p+k}GA^{p+k-1} = A^{p+k-1}$. Similarly, we also have $A^{p+k-1}GA^{p+k} = A^{p+k-1}$. Now, let $H = GA^{p+k-1}G$. Then $A^{p+k+1}HA^{p+k+1} = A^{p+k+1}GA^{p+k-1}GA^{p+k+1}AA^{p+k-1} = GA^{p+k+1} = AA^{p+k-1}A = A^{p+k+1}$. Thus $A^{p+k+1}$ is regular. To prove the second part simply observe that $A^{p+k-1}AGA^{p+k-1} = A^{p+k}GA^{p+k-1} = A^{p+k-1}$. Thus $A^{p+k-1}$ is regular.

Let us show (c). Let $H = A^p(A^{2p+1})^-A^p$, we shall show that $H$ is the Drazin inverse of $A$. Since, $\rho(A^{2p+1}) = \rho(A^{p+1}) = \rho(A^p)$ there exists matrices

$X, Y$ and $Z$ over the field $F$ of quotients of $R$ such that

$$\begin{aligned} A^{p+1} &= XA^{2p+1} \\ A^p &= YA^{2p+1} \\ \text{and} \qquad A^p &= A^{2p+1}Z. \end{aligned}$$

Now, $AH = AA^p(A^{2p+1})^-A^p = XA^{2p+1}(A^{2p+1})^-A^{2p+1}Z = A^{p+1}Z$ and $HA = A^p(A^{2p+1})^-A^pA = YA^{2p+1}(A^{2p+1})^-AA^p = YA^{2p+1}(A^{2p+1})^-AA^{2p+1}Z = YA^{2p+1}AZ = A^{p+1}Z$.

Thus $AH = HA$.

$$A^{p+1}H = A^{p+1}A^p(A^{2p+1})^-A^p = A^{2p+1}(A^{2p+1})^-A^p = A^p.$$

Thus $(1^p)$ is satisfied.

Finally,

$$\begin{aligned} HAH &= A^p(A^{2p+1})^-A^pA^{p+1}Z = A^p(A^{2p+1})^-A^{2p+1}Z \\ &= A^p(A^{2p+1})^-A^p \quad \text{since} \quad A^{2p+1}Z = A^p. \end{aligned}$$

Thus $H$ is the Drazin inverse of $A$.

For a $k \geq 1$, observe that, if $A^{p+k}$ is regular then $A^{2p+1}$ is regular. This follows from (b) by going forwards if $k < p+1$ and going backwards if $k > p+1$. Now, by (c) we get that $A$ has Drazin inverse. This proves (d). ♠

Similar to part (a) of Proposition 6.24 we can also show that $A^p$ is regular if $A$ has Drazin inverse. Indeed, if $A$ has Drazin inverse we have $A^{p+1}G = A^p$. Hence $A^pGA = A^p$. But $GA = G^pA^p$. Hence $A^pG^pA^p = A^p$ or $A^p$ is regular. Not only this, even $A^k$ is regular for every $k \geq p$, because $A^kG^kA^k = A^kGA = A^k$.

However regularity of $A^p$ does not guarantee the existence of Drazin inverse of $A$ as the ensuing example shows.

**Example:** Let $A = \begin{bmatrix} 1 & 1 \\ 1 & 1 \end{bmatrix}$ over the integral domain $\mathbb{Z}$. Then the index of $A = p = 1$. and $A$ is regular over $\mathbb{Z}$. But $A^2 = \begin{bmatrix} 2 & 2 \\ 2 & 2 \end{bmatrix}$ is not regular over $\mathbb{Z}$. Hence $A$ does not have Drazin inverse over $\mathbb{Z}$ because of Proposition 6.24 (a).

But if $A^p$ has a group inverse then $A$ has Drazin inverse as the following Proposition shows.

**Proposition 6.25:** Over an integral domain $R$, an $m \times m$ matrix $A$ has Drazin inverse if and only if $A^p$ has Group inverse, where $p$ is the index of $A$. If $k \geq p$, $A$ has Drazin inverse if and only if $A^k$ has group inverse.

**Proof:** If $G$ is the Drazin inverse of $A$ we have already seen that $A^pG^pA^p = A^p$. Let us see that $G^p$ is the group inverse of $A^p$. $G^pA^pG^p = G^pAG = G^{p-1}G = G^p$. Also $A^pG^p = AG = GA = G^pA^p$.

Conversely, if $A^p$ has group inverse then by Theorem 6.19 (v) $A^{2p}$ is regular and by Proposition 6.24 (b) $A^{2p+1}$ is regular and by Proposition 6.24 (c) we have that $A$ has Drazin inverse. The second part is proved similarly.

♠

Now we shall go in a different direction – namely – the compound matrices.

**Theorem 6.26:** Let $A$ be an $m \times m$ matrix over an integral domain $R$ with index $p$. Let $\rho(A^p) = s$. Then the following are equivalent.

(i) $A$ has Drazin inverse.

(ii) $C_s(A)$ has Drazin inverse.

(iii) $C_s(A^p)$ has Group inverse.

(iv) $Tr(C_s(A^p))$ is a unit.

(v) $A^p$ has group inverse.

**Proof:** If $s = 0$ there is really nothing to show. Let $s \neq 0$. (i) $\Rightarrow$ (ii) follows from the properties of compound matrices.

(ii) $\Rightarrow$ (iii). Since $\rho(A^p) = s = \rho(A^{p+1})$ and $s \neq 0$, $\rho([C_s(A)]^p) = \rho(C_s(A^p)) = 1 = \rho(C_s(A^{p+1})) = \rho([C_s(A)]^{p+1})$. Thus index of $C_s(A)$ is $\leq p$. If index of $C_s(A) = q$ then $q \leq p$ and since $C_s(A)$ has Drazin inverse, by Proposition 6.25, $[C_s(A)]^p$ has group inverse.

(iii) $\Rightarrow$ (iv) follows from (ii) $\Rightarrow$ (iii) of Theorem 6.20.

(iv) $\Rightarrow$ (v) follows from (iii) $\Rightarrow$ (i) of Theorem 6.20.

In both the above implications Theorem 6.20 is applicable since $\rho(A^p) = s$.

(v) $\Rightarrow$ (i). Since $A^p$ has group inverse, from Theorem 6.19 we get that $\rho(A^p) = \rho(A^{2p})$ and $A^{2p}$ is regular. Since $p \geq 1$, from Proposition 6.24 (d) we get that $A$ has Drazin inverse.

Thus Theorem 6.26 is proved.

♠

Let us also observe that the Drazin inverse of $A$ when it exists is a polynomial in $A$.

**Proposition 6.27:** For an $m \times m$ matrix $A$ over an integral domain $R$, if $A$ has Drazin inverse, the Drazin inverse of $A$ is a polynomial in $A$.

**Proof:** Let the index of $A$ be $p$. From Theorem 6.24 (a) and (b), since $A$ has Drazin inverse, we get that $A^{2p+1}$ is regular.

If we write $B = A^{2p+1}$ then $\rho(B) = \rho(B^2) = \rho(A^p)$. And again from Theorem 6.24 (b) we have that $B^2$ is regular. Hence by Theorem 6.19 (i) $\Leftrightarrow$ (v) we get that $B$ has group inverse.

From the final statement of Theorem 6.24 we have that $A^p(A^{2p+1})^- A^p$ is the Drazin inverse of $A$, where $(A^{2p+1})^-$ is any $g$-inverse of $A$. If we take $(A^{2p+1})^- = (A^{2p+1})^\#$ then $(A^{2p+1})^\#$ and hence $A^p(A^{2p+1})^\# A^p$ is a polynomial in $A$ from Theorem 6.21. Hence the Proposition.

♠

If a matrix $A$ has group inverse then $p = 1$ and $A$ has Drazin inverse. If the index of a matrix $A$ is $\neq 1$ then $A$ does not have group inverse, but, it may still have Drazin inverse. We shall now prove a (unique) decomposition Theorem writing any matrix having a Drazin inverse, as a sum of two matrices one of which has group inverse.

**Theorem 6.28:** Let $A$ be an $m \times m$ matrix over an integral domain $R$. If $A$ has Drazin inverse then $A$ can be written as $A_1 + A_2$ where (i) $A_1$ has group inverse (ii) $A_2$ is nilpotent and (iii) $A_1A_2 = A_2A_1 = 0$. Such a decomposition is also unique. Conversely, every matrix $A$ which can be written as $A_1 + A_2$ where $A_1$ and $A_2$ satisfy (i), (ii) and (iii) has Drazin inverse.

**Proof:** Let $K$ be the Drazin inverse of $A$. Let the index of $A$ be $= p$.

Then $KAK = K, KA = AK$ and $A^{p+1}K = A^p$. Hence $K$ has a commuting $g$-inverse, namely $A$. This implies that $AKA$ is the group inverse of $K$. Let $A_1 = AKA = K^\#$. Let $A_2 = A - A_1$. Then $A = A_1 + A_2$ and $A_1$ has group inverse (namely $K$). We shall show that $A_1A_2 = A_2A_1 = 0$ and $A_2^p = 0$.

Since $KAK = K$ we have $AKAK = AK$. Also we have $AK = KA$. Hence $A_1A_2 = AKA(A - AKA) = AKA^2 - AKAAKA = AKA^2 - AKAKA^2 = AKA^2 - AKA^2 = 0$.

Also $A_2A_1 = (A - AKA)AKA = A^2KA - A^2KA = 0$. Now, $A_2^p = (A-AKA)^p = [A(I-KA)]^p = A^p(I-AK)(I-AK)^{p-1} = (A^p - A^{p+1}K)(I-AK)^{p-1} = 0$.

Thus $A_2$ is nilpotent. We have proved the decomposition.

To prove the uniqueness let $A = B_1 + B_2$ where $B_1$ has group inverse, $B_2$ is nilpotent and $B_1B_2 = B_2B_1 = 0$.

Let $B_1^\#$ be the group inverse of $B_1$. Then

$$(B_1^\#)^2 B_1 = B_1(B_1^\#)^2 = B_1^\#.$$

This gives us that $B_2B_1^{\#} = B_2B_1(B_1^{\#})^2 = 0$ and also $B_1^{\#}B_2 = 0$. Hence $AB_1^{\#} = (B_1 + B_2)B_1^{\#} = B_1B_1^{\#} + B_2B_1^{\#} = B_1B_1^{\#}$ and we also get that $B_1^{\#}A = B_1^{\#}B_1 = B_1B_1^{\#}$. Hence we have $A(B_1^{\#})^2 = B_1(B_1^{\#})^2 = B_1^{\#}$.

If $\ell \geq 1$ is such that $B_2^{\ell} = 0$ then $A^{\ell} = B_1^{\ell}$. From this we get that $A^{\ell+1}B_1^{\#} = B_1^{\ell-1}B_1^2B_1^{\#} = B_1^{\ell-1}B_1 = B_1^{\ell} = A^{\ell}$. Thus we have that $A^{\ell+1}B_1^{\#} = A^{\ell}, AB_1^{\#} = B_1^{\#}A$ and $A(B_1^{\#})^2 = B_1^{\#}$. Thus $B_1^{\#}$ is the Drazin inverse of $A$. This actually means that $B_1$ is the group inverse of the Drazin inverse of $A$. Hence $B_1$ is uniquely defined. Hence the decomposition is unique.

In the above proof of uniqueness we have also shown that if $A$ has a decomposition satisfying (i), (ii) and (iii) then $A$ has Drazin inverse.

♠

## EXERCISES

**Exercise 6.1:** Give an example to show that (i) of Theorem 6.19 is not equivalent to saying that $\rho(A) = \rho(A^2)$ and $A$ is regular.

**Exercise 6.2:** State and prove the analogues of Theorem 6.16, Theorem 6.17 and Proposition 6.18 for the Moore-Penrose inverse of a matrix $A$.

**Exercise 6.3:** For an $m \times n$ matrix $A$ over an integral domain with an involution $a \to \bar{a}$ $a \to \bar{a}$, show that $A$ has Moore-Penrose inverse if and only if $\rho(A^*AA^*) = \rho(A)$ has group $A^*AA^*$ is regular.

**Exercise 6.4:** For an $m \times n$ matrix $A$ over an integral domain show that $A$ has Moore-Penrose inverse if and only if $A^*A$ has group inverse and $\rho(A^*A) = \rho(A)$ (equivalently $AA^*$ has group inverse and $\rho(AA^*) = \rho(A)$).

**Exercise 6.5:** (a) For a matrix $A$ over an integral domain $R$, show that $\rho(A) = \rho(A^2)$ if and only if $\sum_{\gamma \in Q_{r,m}} |A_{\gamma\gamma}| \neq 0$ where $r = \rho(A)$.

(b) Does the result of (a) hold for matrices over commutative rings?

**Exercise 6.6:** Over an integral domain, if a matrix $A$ has Drazin inverse, give an explicit polynomial expression for the Drazin inverse of $A$.

**Exercise 6.7:** Over any associative ring, if a matrix $A$ has Drazin inverse, show that it must be unique. More generally, in any associative ring define Drazin inverse of an element in the ring and show that the Drazin inverse of an element is unique whenever it exists.

**Exercise 6.8:** Over an integral domain with an involution $a \to \bar{a}$, let $M$ and $N$ be invertible matrices. Following [69] we say that a matrix $G$ is a *generalized Moore-Penrose inverse (with respect to M and N)* of a matrix $A$ if $A$ and $G$ satisfy

$$AGA = A \tag{1}$$

$$GAG = G \tag{2}$$

$$(MAG)^* = MAG \tag{3}$$

$$(NGA)^* = NGA \tag{4}$$

Find necessary and sufficient conditions for a matrix to admit generalized Moore-Penrose inverse with respect to $M$ and $N$.

# Chapter 7

# Regularity – commutative rings

## 7.1 Commutative rings with zero divisors

We have seen many examples of integral domains and investigated regularity of matrices over them. Let us now see some examples of commutative rings with zero divisors and investigate regularity of matrices over them.

**Example 1:** Let $\mathbb{Z}[k]$ be the ring of integers modulo $k$. Then

(i) an element $\ell \in \mathbb{Z}[k]$ is regular if and only if $(\ell^2, k)|\ell$;

(ii) a matrix $A$ over $\mathbb{Z}[k]$ has a $g$-inverse if and only if all the invariant factors of $A$ modulo $k$ have $g$-inverses in $\mathbb{Z}[k]$; and

(iii) every matrix $A$ over $\mathbb{Z}[k]$ has a $g$-inverse if and only if $k$ is square free.

All the above statements are easily established. For example, in the ring $\mathbb{Z}[12]$, $0, 1, 3, 4, 5, 7, 8, 9, 11$ are regular and $2, 6, 10$ are not regular. The ring $\mathbb{Z}[12]$ is useful for constructing many counterexamples.

**Example 2:** Let $R_0$ be the ring of all real valued continuous functions on the real line. Note that, in this ring the only idempotents are 0 and 1. Hence (i)–(vi) of Theorem 5.3 are all equivalent. For a matrix $A$ over $R_0$ and a real number $x$, let $A(x)$ be the real matrix whose $(i, j)$th entry is $a_{ij}(x)$.

For a matrix $A$ over $R_0$ with $\rho(A) = r > 0$ the following are equivalent.

(i) $A$ is regular.

(ii) $\rho(A(x)) = r$ for every $x$.

(iii) The $r \times r$ minors of $A$ have no common zero, i.e., there is no $x$ such that $|A_{\alpha\beta}|(x) = 0$ for all $(\alpha, \beta) \in \Delta$.

(iv) A linear combination of all the $r \times r$ minors of $A$ with coefficients from $R_0$ is equal to one.

If $G$ is a $g$-inverse of $A$ then $x \to \rho(A(x)) = \rho(AG(x)) = Tr(AG)(x)$ is a continuous function of $x$. Hence, since the real line is connected, $\rho(A(x)) = r$ for all $x$. Thus (i) $\Rightarrow$ (ii) is proved. (ii) $\Rightarrow$ (iii) is clear.

(iii) $\Rightarrow$ (iv) follows from topological considerations and since these considerations are not so standard we shall give the details. Let $f_1, f_2, \cdots, f_k$ be the real valued continuous functions with no common zero. Let $Z_i = \{x : f_i(x) = 0\}$ for $i = 1, 2, \cdots, k$. Since $f_1, f_2, \cdots, f_k$ have no common zero, $\bigcap_{i=1}^{k} Z_i = \emptyset$. By a result from general topology, viz., ([33], p. 266, Theorem 4) there exist open sets $V_i$ for $i = 1, 2, \cdots, k$ such that $Z_i \subset V_i$ for $i = 1, 2, \cdots, k$ and $\bigcap_{i=1}^{k} V_i = \emptyset$. By the normality of the real line there exist open sets $W_i$ for $i = 1, 2, \cdots, k$ such that $Z_i \subset W_i \subset \overline{W_i} \subset V_i$ for $i = 1, 2, \cdots, k$. By Urysohn's lemma we get real valued continuous functions $h_i$ taking values in $[0, 1]$ such that $h_i(x) = 0$ if $x \in \overline{W_i}$ and $h_i(x) = 1$ if $x \notin V_i$ for $i = 1, 2, \cdots, k$. Then the function $h$ defined by $h(x) = \sum_{i=1}^{k} h_i(x)$ never vanishes since $\bigcap_{i=1}^{k} V_i = \emptyset$. Also for $i = 1, 2, \cdots, k$ the function $g_i(x) = \frac{h_i(x)}{h(x)f_i(x)}$ is a continuous function since on the open set $W_i$, $g_i(x) = 0$ and on the open set $Z_i^c, h(x)f_i(x) \neq 0$. Clearly, $\sum_{i=1}^{k} f_i g_i = 1$.

(iv) $\Rightarrow$ (i) was already seen in Theorem 5.3.

A corresponding result can be obtained for matrices over the ring of all real valued continuous functions over a normal topological space. We shall give this as an Exercise at the end of the Chapter. The following refinement of the above example would be required for the solution.

**Example 3:** Let $R_0$ be the ring of all real valued continuous functions on some normal topological space X. Note that, in this ring there may exist idempotents other than 0 and 1. Hence (i) need not imply (ii) of the above example. However, for any matrix $A$ over $R_0$ (ii), (iii) and (iv) of the above example are equivalent.

## 7.2 Rank one matrices

For the study of generalized inverses of matrices over commutative rings we are going to make use of the matrix $C_r(A)$ which is of rank one if $\rho(A) = r$.

In view of this we shall study rank one matrices over commutative rings in relation to regularity. Our first result contains a generalization of Theorem 5.2.

**Proposition 7.1:** Let $A, G$ and $B$ be $m \times n, n \times m$ and $m \times n$ matrices respectively over a commutative ring $R$. Let further $\rho(A) = 1$. Then $AGA = Tr(AG)A$. In particular $AGA = B$ if and only if $Tr(AG)A = B$.

The proof of this Proposition is contained in the proof of Theorem 5.1. ♠

We shall see below several results that follow from this Proposition. Only one part of most of the following results will be used later.

**Theorem 7.2:** Let $R$ be a commutative ring and $A$ be a matrix over $R$ with $\rho(A) = 1$. Let $G$ be a matrix over $R$. Then

(a) $G$ is a $g$-inverse of $A$ if and only if $Tr(AG)A = A$.

(b) $G$ is a reflexive $g$-inverse of $A$ if and only if $\rho(G) = 1, Tr(AG)A = A$ and $Tr(AG)G = G$.

**Proof:** (a) follows from Proposition 7.1. To prove (b) observe that if $AGA = A$ and $\rho(A) = 1$ then $G \neq 0$. Also, $GAG = G$ tells us that $\rho(G) \leq \rho(A)$. Hence $\rho(G) = 1$. All the other parts of (b) follow from Proposition 7.1. ♠

Characterizing rank one matrices which admit Moore-Penrose inverses is also not difficult.

**Theorem 7.3:** Let $R$ be a commutative ring with an involution $a \to \bar{a}$ and let $A$ be a matrix over $R$ with $\rho(A) = 1$. Then $G$ is the Moore-Penrose inverse of $A$ if and only if $Tr(A^*A)G = A^*$ and $Tr(G^*G)A = G^*$.

Further, if $G$ is the Moore-Penrose inverse of $A$ then $Tr(A^*A)Tr(G^*G)A = A$ and the element $Tr(G^*G)$ is the Moore-Penrose inverse of $Tr(A^*A)$.

**Proof:** Let $A$ and $G$ satisfy $AGA = A, GAG = G, (AG)^* = AG$ and $(GA)^* = GA$. Then $G^*GA = G^*A^*G^* = G^*$ and also $AGG^* = G^*$. Now, $AGG^*GA = AGG^* = G^*$. By Proposition 7.1 we have $Tr(AGG^*G)A = G^*$. But $AGG^*G = G^*G$. Hence $Tr(G^*G)A = G^*$. Similarly we also have $Tr(A^*A)G = A^*$.

Now, observe that $\overline{Tr(A^*A)} = \overline{\sum_i \sum_k \overline{a_{ki}} a_{ki}} = \sum_i \sum_k a_{ki}\overline{a_{ki}} = Tr(A^*A)$. Also observe that for any element $a$ of $R$ and matrix $C$, $(aC)^* = \bar{a}C^*$.

Let $A$ and $G$ satisfy $Tr(G^*G)A = G^*$ and $Tr(A^*A)G = A^*$. Then from the observations above we get that $Tr(A^*A)G^* = A$. Hence $Tr(A^*A)$ $Tr(G^*G)A = A$. Also we have,

$$Tr(A^*A)Tr(G^*G)A^*A = A^*A.$$

Hence $Tr(A^*A)Tr(G^*G)Tr(A^*A) = Tr(A^*A)$. From, $Tr(A^*A)Tr(G^*G)A = A$ and $GAG = G$ we get that,

$$Tr(A^*A)Tr(G^*G)G = G.$$

Hence we have $Tr(G^*G)Tr(A^*A)Tr(G^*G) = Tr(G^*G)$. In any case,

$$(Tr(A^*A)Tr(G^*G))^* = Tr(A^*A)Tr(G^*G).$$

Thus, $Tr(G^*G)$ is the Moore-Penrose inverse of $Tr(A^*A)$. Thus the last part of the Theorem is proved.

Let us now suppose that $Tr(A^*A)G = A^*$ and $Tr(G^*G)A = G^*$. We shall show that $A$ and $G$ satisfy the Moore-Penrose equations. From what we have seen in the previous paragraph we get that $Tr(AA^*)Tr(G^*G)A = Tr(A^*A)Tr(G^*G)A = A$. Hence, by Proposition 7.1 we get $Tr(G^*G)AA^*A = A$. But $Tr(G^*G)A^* = G$. Thus we have $AGA = A$ and $Tr(A^*A)Tr(G^*G)A^* = A^*$. Again by Proposition 7.1 we get that $Tr(G^*G)A^*AA^* = A^*$. Thus $GAG = G$. Also, $(AG)^* = Tr(G^*G)AA^* = AG$. Similarly we have $(GA)^* = GA$. Thus $G$ is the Moore-Penrose inverse of $A$.

♠

If in addition to $\rho(A) = 1$, $A$ is also regular then, nice necessary and sufficient conditions can be given for $A$ to admit Moore-Penrose inverse.

**Proposition 7.4:** Let $A$ be a matrix over a commutative ring $R$ with an involution $a \to \bar{a}$ with $\rho(A) = 1$. If $A$ is regular with a $g$-inverse $G$ then $A$ has Moore-Penrose inverse if and only if $\overline{Tr(AG)} = Tr(AG)$ and $Tr(A^*A)|Tr(AG)$. If $w$ is an element such that $Tr(A^*A)w = Tr(AG)$ and $\bar{w} = w$ then $wA^*$ is the Moore-Penrose inverse of $A$.

**Proof:** If $U$ and $H$ are two $g$-inverses of $A$ then $Tr(AU) = Tr(AHAU) = Tr(AUAH) = Tr(AH)$. If $A^+$ is the Moore-Penrose inverse of $A$, we see that $Tr(AG) = Tr(AA^+)$. But $(AA^+)^* = AA^+$. Hence $\overline{Tr(AA^+)} = Tr(AA^+)$. From Theorem 7.3 we have $Tr(A^{+*}A^+)A^* = A^+$. Hence $Tr(A^{+*}A^+)Tr(AA^*) = Tr(AA^+)$. Thus $Tr(AA^*)|Tr(AA^+)$. But $Tr(AA^+) = Tr(AG)$. Thus we have $Tr(AA^*)|Tr(AG)$.

To show the converse, let $Tr(A^*A)u = Tr(AG)$. Since $\rho(AG) = 1$ and $(AG)^2 = AG$ by Proposition 7.1 we have $Tr(AG)Tr(AG) = Tr(AG)$. Since $\overline{Tr(AG)} = Tr(AG)$ and $\overline{Tr(A^*A)} = Tr(A^*A)$ we have $Tr(A^*A)\bar{u} = Tr(AG)$. Combining the two equalities we get that $Tr(A^*A)uTr(A^*A)\bar{u} = Tr(AG)$. If we let $w = uTr(A^*A)\bar{u}$ then $\bar{w} = w$ and $Tr(A^*A)w = Tr(AG)$.

For this $w$ let $H = wA^*$. Let us show that $H$ is the Moore-Penrose inverse of $A$. Since $\rho(A) = 1$ we have $AA^*A = Tr(AA^*)A$. Hence $AHA = wAA^*A = wTr(AA^*)A = Tr(AG)A = AGA = A$. Since $\rho(A^*) = 1$ we have $A^*AA^* = Tr(A^*A)A^*$. Hence $HAH = w^2A^*AA^* = w^2Tr(A^*A)A^* = wTr(AG)A^* = w\overline{Tr(AG)}A^* = wTr(A^*G^*)A^* = wA^*G^*A^* = wA^* = H$. Also $(AH)^* = (wAA^*)^* = \bar{w}AA^* = wAA^* = AH$. Similarly we also have $(HA)^* = HA$. Thus $H = A^+$.

♠

We shall now look at Group inverses of rank one matrices.

**Theorem 7.5:** Let $R$ be a commutative ring and $A$ and $G$ be $n \times n$ matrices over $R$ with $\rho(A) = 1$.

(a) If $G$ is the group inverse of $A$ then $Tr(G)A = GA, Tr(A)G = AG$, $Tr(G)Tr(A) = Tr(GA), (Tr(G))^2A = G$ and $(Tr(A))^2G = A$. More generally, $Tr(G^kA^\ell) = (Tr(G))^k(Tr(A))^\ell$ for all $k, \ell \geq 1$.

(b) Conversely, if $(Tr(G))^2A = G$ and $(Tr(A))^2G = A$ then $G$ is the group inverse of $A$.

**Proof:** If $G$ is the group inverse of $A$ then $AGA = A, GAG = G$ and $AG = GA$ hold. Note that $\rho(G) = 1$. Hence, $GA^2G = AGAG = AG$. From Proposition 7.1 we get that $Tr(GA^2)G = AG$. But $GA^2 = A$. Thus $Tr(A)G = AG$. Similarly we get $Tr(G)A = GA$. Also we get $Tr(G)Tr(A) = Tr(GA)$. From one of the Exercises at the end of this Chapter we also get that $Tr(A^k) = (Tr(A))^k$. Hence using $Tr(G)A = GA$ and $Tr(A)G = AG$ and $AG = GA$ we have for all $k$ and $\ell \geq 1, Tr(G^kA^\ell) = (Tr(G))^k(Tr(A))^\ell$. Also $(Tr(G))^2A = Tr(G)GA = Tr(G)AG = GAG = G$. Similarly $(Tr(A))^2G = AGA = A$.

We shall now show the converse.

Let $(Tr(G))^2A = G$ and $(Tr(A))^2G = A$. Note that $\rho(G) = 1$. Clearly $AG = (Tr(G))^2A^2 = GA$. Note also that since $\rho(A) = 1$, $A^3 = (Tr(A))^2A$. Now, $AGA = (Tr(G))^2A^3 = (Tr(G))^2(Tr(A))^2A = (Tr(A))^2G = A$. Also, $GAG = (Tr(G))^4A^3 = (Tr(G))^2A = G$. Thus $G$ is the group inverse of $A$.

♠

If in addition to $\rho(A) = 1$, $A$ is also regular, nice necessary and sufficient conditions can be given for $A$ to admit group inverse.

**Theorem 7.6:** Let $R$ be a commutative ring and $A$ be an $n \times n$ matrix over $R$ with $\rho(A) = 1$. If $A$ is regular with a $g$-inverse $G$ then, $A$ has group inverse if and only if $(Tr(A))^2 | Tr(AG)$. If $w$ is an element such that $(Tr(A))^2w = Tr(AG)$ then $wA$ is the group inverse of $A$.

**Proof:** If $A^\#$ is the group inverse of $A$, as in the beginning of the proof of Theorem 7.4 we have $Tr(AA^\#) = Tr(AG)$. Hence $Tr(AA^\#) = Tr(AA^\#AA^\#) = Tr(A^2A^{\#^2}) = (Tr(A^2))(Tr(A^{\#^2})) = (Tr(A))^2(Tr(A^\#))^2$. Thus $(Tr(A))^2 | Tr(AG)$.

Conversely, let $(Tr(A))^2w = Tr(AG)$. Let $H = wA$. Then $AHA = wA^3 = w(Tr(A))^2A = Tr(AG)A = AGA = A$. Also, $HAH = w^2A^3 = wA = H$. That, $AH = HA$ is clear. Thus $H$ is the group inverse of $A$.

♠

In the next section using the above results we shall characterize some special types of matrices which are regular.

## 7.3 Rao-regular matrices

Let $A$ be an $m \times n$ matrix over a commutative ring with $\rho(A) = r$. Recall from section 4.2 that the ideals $\mathcal{I}_0(A), \mathcal{I}_1(A), \mathcal{I}_2(A), \cdots, \mathcal{I}_k(A), \cdots$ are defined by $\mathcal{I}_0(A) = R$, for $1 \leq k \leq \min(m,n)$ $\mathcal{I}_k(A) =$ the ideal generated by all the $k \times k$ minors of $A$ and $\mathcal{I}_k(A) = \{0\}$ for all other $k$. In particular, $\mathcal{I}_0(A) \supset \mathcal{I}_1(A) \supset \mathcal{I}_2(A), \cdots$, $\{0\} = \mathcal{I}_{r+1}(A) = \mathcal{I}_{r+2}(A) = \cdots$ and $\mathcal{I}_r(A) \neq \{0\}$.

Let us also recall from Theorem 5.3 (ii) ⇒ (iii) that if there exist elements $\{c_{\beta\alpha} : (\alpha,\beta) \in \Delta\}$ in $R$ such that $\left(\sum_{(\alpha,\beta)\in\Delta} |A_{\alpha\beta}|c_{\beta\alpha}\right) a_{k\ell} = a_{k\ell}$ for all $k$ and $\ell$ then $A$ is regular. Theorem 5.3 (iii) ⇒ (iv) ⇒ (v) tells us that if $A$ is regular then there exist elements $\{c_{\beta\alpha} : (\alpha,\beta) \in \Delta\}$ in $R$ such that $\left(\sum_{(\alpha,\beta)\in\Delta} |A_{\alpha\beta}|c_{\beta\alpha}\right) |A_{\gamma\delta}| = |A_{\gamma\delta}|$ for all $(\gamma,\delta) \in \Delta$. Using the notion of the ideals $\mathcal{I}_k(A)$ for $k \geq 0$ we can reformulate these results as

**Proposition 7.7:** Let $A$ be an $m \times n$ matrix over a commutative ring $R$ such that $\rho(A) = r$.

(a) If $A$ is regular then there is an element $e \in \mathcal{I}_r(A)$ such that $e$ is the identity of the ring $\mathcal{I}_r(A)$ (equivalently, $e|A_{\gamma\delta}| = |A_{\gamma\delta}|$ for all $(\gamma,\delta) \in \Delta$ or $eC_r(A) = C_r(A)$).

(b) If there is an element $e \in \mathcal{I}_r(A)$ such that $eA = A$ (equivalently, $e$ is an identity of $\mathcal{I}_1(A)$ though $e$ may not be an element of $\mathcal{I}_1(A)$), then $A$ is regular.

If an element $e$ of $R$ is as in (b) then $e$ satisfies (a) also. If $e$ is as in (a) then $e$ is idempotent.

**Proof:** (a) and (b) are reformulations of Theorem 5.3 (ii) ⇒ (iii) ⇒ (iv) ⇒ (v). Clearly, if $eA = A$ then $eC_r(A) = C_r(A)$.

♠

Following [88] and [71], taking (b) of the previous Proposition as the basis we shall call a matrix $A$ over a commutative ring $R$ with $\rho(A) = r$, *Rao-regular* if there exists an $e \in \mathcal{I}_r(A)$ such that $eA = A$. The idempotent element $e$ will be called the *Rao-idempotent* of $A$. Let us first see that this element $e$ is uniquely defined.

**Proposition 7.8:** Let $A$ be an $m \times n$ matrix over a commutative ring $R$ with $\rho(A) = r$.

(a) $e \in \mathcal{I}_r(A)$ such that $eC_r(A) = C_r(A)$, if exists, is unique.

(b) $e \in \mathcal{I}_r(A)$ such that $eA = A$, if exists, is unique.

**Proof:** Observe that $\mathcal{I}_r(A)$ in its own right is a ring. If $e \in \mathcal{I}_r(A)$ is such that $eC_r(A) = C_r(A)$ then $e$ is the identity of $\mathcal{I}_r(A)$. Hence such an $e$, if exists is unique.

If $eA = A$ then $eC_r(A) = C_r(A)$. Hence $e \in \mathcal{I}_r(A)$ such that $eA = A$, if exists, is unique.

♠

We shall write $I(A)$ for the Rao-idempotent of a Rao-regular matrix $A$. Thus, if $\rho(A) = r, I(A)$ is the element $e$ of $\mathcal{I}_r(A)$ such that $eA = A$. Let us first observe some properties of Rao-regular matrices.

**Theorem 7.9:** Consider matrices over a commutative ring $R$.

(a) Every Rao-regular matrix is regular. In fact, if $A$ is an $m \times n$ Rao-regular matrix over $R$ with $\rho(A) = r$ and with Rao-idempotent $I(A) = \sum_{(\alpha,\beta)\in\Delta_{r,m,n}} |A_{\alpha\beta}|c_{\beta\alpha}$ then the $n \times m$ matrix $G = (g_{ij})$ defined by

$$g_{ij} = \sum_{(\alpha,\beta)\in\Delta} \frac{\partial |A_{\alpha\beta}|}{\partial a_{ji}} c_{\beta\alpha}$$

is a $g$-inverse of $A$.

(b) If 0 and 1 are the only idempotents of $R$ then every regular matrix is Rao-regular.

(c) Over a general commutative ring not every regular matrix need be Rao-regular.

(d) Every regular matrix of rank one is Rao-regular.

**Proof:** The proof of (a) is really the proof of (ii) $\Rightarrow$ (iii) of Theorem 5.3.

(b) follows from Theorem 5.3.

For (c) consider the matrix $A = \begin{bmatrix} 2 & 0 & 0 \\ 0 & 2 & 0 \\ 0 & 0 & 3 \end{bmatrix}$ over $\mathbb{Z}[6]$. Then $\rho(A) = 2$ and $G = \begin{bmatrix} 2 & 0 & 0 \\ 0 & 2 & 0 \\ 0 & 0 & 1 \end{bmatrix}$ is a $g$-inverse of $A$. But $C_2(A) = \begin{bmatrix} 4 & 0 & 0 \\ 0 & 0 & 0 \\ 0 & 0 & 0 \end{bmatrix}$. Whatever $\sum_{(\alpha,\beta)\in\Delta} |A_{\alpha\beta}|c_{\beta\alpha}$ we take, $\sum_{(\alpha,\beta)\in\Delta} |A_{\alpha\beta}|c_{\beta\alpha}$ will be of the form $4a$

for some $a \in \mathbb{Z}[6]$ and $4aA = A$ is not possible. Thus $A$ is regular but not Rao-regular.

(d) is clear.

♠

Though not every regular matrix is Rao-regular, from a regular matrix we can extract a Rao-regular component as in the following Proposition. Some times even from any matrix also a Rao-regular component can be extracted.

**Proposition 7.10:** Let $A$ be an $m \times n$ matrix over a commutative ring $R$ with $\rho(A) = r > 0$. If there is an identity element $e$ of $\mathcal{I}_r(A)$, then, $\rho(eA) = r$, $eA$ is Rao-regular $I(eA) = e$ and $\rho(A - eA) < r$. In case $A$ is regular there is an identity element $e$ of $\mathcal{I}_r(A)$ and it also has the property that $A - eA$ is regular.

**Proof:** Let $e \in \mathcal{I}_r(A)$ be an identity element of $\mathcal{I}_r(A)$, i.e., $e|A_{\gamma\delta}| = |A_{\gamma\delta}|$ for all $(\gamma, \delta) \in \Delta$. If we consider $C_r(eA)$, since $e$ is idempotent we have $C_r(eA) = eC_r(A) = C_r(A)$. Thus $C_r(eA) = C_r(A) \neq 0$. Also $C_{r+1}(eA) = C_{r+1}(A) = 0$. Hence $\rho(eA) = r$. Also the ideal $\mathcal{I}_r(eA) = \mathcal{I}_r(A)$ since $C_r(eA) = C_r(A)$. Hence $e \in \mathcal{I}_r(eA)$ and $e$ is the identity element of $\mathcal{I}_r(eA)$. Also, $eeA = eA$ since $e$ is idempotent. Thus $eA$ is Rao-regular and $I(eA) = e$. $C_r((1-e)A) = (1-e)C_r(A)$ because $1 - e$ is also idempotent. But $eC_r(A) = C_r(A)$. Hence $C_r((1-e)A) = 0$. This shows that $\rho((1-e)A) < r$. In case $A$ is regular, by Proposition 7.7 (a) there is an identity element $e$ in $\mathcal{I}_r(A)$. Also, if $G$ is a $g$-inverse of $A$ then $G$ is also a $g$-inverse of $A - eA$.

♠

This Proposition will be the basis of our main Theorem of the next section.

We shall now establish some more properties of Rao-regular matrices and their Rao-idempotents.

**Proposition 7.11:** Let $A$ be an $m \times n$ Rao-regular matrix over a commutative ring $R$ with $\rho(A) = r > 0$ and Rao-idempotent $I(A)$. Then,

(a) $C_r(A)$ is Rao-regular and $I(C_r(A)) = I(A)$.

(b) If $G$ is a $g$-inverse of $A$ then $AG$ and $GA$ are also Rao-regular, $\rho(AG) = \rho(GA) = \rho(A)$ and $I(AG) = I(GA) = I(A)$.

(c) If $G$ is a reflexive $g$-inverse of $A$ then $\rho(G) = r, G$ is Rao-regular and $I(G) = I(A)$.

**Proof:** Since $A$ is Rao-regular by Theorem 7.9 (a), $A$ is regular. Hence $C_r(A)$ is regular. But $\rho(C_r(A)) = 1$. By Theorem 7.9 (d) we have that $C_r(A)$ is Rao-regular. Also, since $I(A)A = A$ we have $I(A)C_r(A) = C_r(A)$.

Since $I(A) \in \mathcal{I}_r(A) = \mathcal{I}_1(C_r(A))$, by the uniqueness of the Rao-idempotent (Theorem 7.8) we have that $I(A) = I(C_r(A))$. Thus (a) is proved.

If $G$ is a $g$-inverse of $A$ and $A$ is Rao-regular with Rao-idempotent $I(A)$, since $I(A)A = A$, it follows that $I(A)AG = AG$. Also, $\mathcal{I}_r(A) \subseteq \mathcal{I}_r(AGA) \subset \mathcal{I}_r(AG) \subset \mathcal{I}_r(A)$ by Proposition 4.6. Thus $I(A) \in \mathcal{I}_r(AG)$, $\rho(AG) = r$ and $I(A)AG = AG$. Hence $AG$ is Rao-regular and $I(AG) = I(A)$. Similarly one can show that $I(GA) = I(A)$. Thus (b) is proved.

If $G$ is a reflexive $g$-inverse of $A$, by Theorem 3.4 we have that $\rho(G) = \rho(A) = r$. From $I(A)A = A$ we also obtain that $I(A)G = I(A)GAG = GAG = G$. Also $\mathcal{I}_r(G) = \mathcal{I}_r(GAG) \subset \mathcal{I}_r(A)$ and $\mathcal{I}_r(A) = \mathcal{I}_r(AGA) \subset \mathcal{I}_r(G)$. Hence $I(A) \in \mathcal{I}_r(A) = \mathcal{I}_r(G)$. Thus $I(A) \in \mathcal{I}_r(G)$ and $I(A)G = G$. Thus $G$ is Rao-regular and $I(G) = I(A)$ by the uniqueness result of Proposition 7.8 (b). Thus (c) is proved.

♠

If $e_1, e_2, \cdots, e_\ell$ are idempotents in R we shall say that they are pairwise orthogonal if $e_ie_j = 0$ whenever $1 \leq i, j \leq \ell$. If $A_1, A_2, \cdots, A_\ell$ are $m \times n$ Rao-regular matrices with Rao-idempotents $I(A_1), I(A_2), \cdots, I(A_\ell)$ we shall say that $A_1, A_2, \cdots, A_\ell$ have pairwise orthogonal Rao-idempotents if $I(A_1), I(A_2), \cdots, I(A_\ell)$ are pairwise orthogonal. We shall investigate the properties of a sum of Rao-regular matrices with pairwise orthogonal Rao-idempotents.

**Proposition 7.12:** If $C$ and $D$ are $m \times n$ matrices and $e$ and $f$ are idempotents such that $ef = 0$ then for any $k \geq 1$ $C_k(eC + fD) = C_k(eC) + C_k(fD)$ and $\mathcal{I}_k(eC + fD) = \mathcal{I}_k(eC) + \mathcal{I}_k(fD)$. Here for ideals $\mathcal{I}$ and $\mathcal{J}, \mathcal{I} + \mathcal{J}$ is the ideal generated by $\mathcal{I} \cup \mathcal{J}$ .

Also, if $A$ and $B$ are $m \times n$ Rao-regular matrices over a commutative ring with pairwise orthogonal Rao-idempotents then,

(a) $C_k(A + B) = C_k(A) + C_k(B)$ and $\mathcal{I}_k(A + B) = \mathcal{I}_k(A) + \mathcal{I}_k(B)$ for all $k \geq 1$.

(b) If $r = \rho(A) = \rho(B)$ then $\rho(A + B) = r, A + B$ is Rao-regular and $I(A + B) = I(A) + I(B)$.

(c) If $r = \rho(A) \geq \rho(B)$ then $A + B$ is regular.

**Proof:** If $C = (c_{ij})$ and $D = (d_{ij})$, and $ef = 0$ then $efd_{k\ell} = 0$ and $fec_{ij} = 0$ for all $k, \ell, i$ and $j$. Also $ec_{ij}fd_{k\ell} = 0$ for all $i, j, k$ and $\ell$. This gives us that $|(eC + fD)_{\gamma\delta}| = |eC_{\gamma\delta}| + |fD_{\gamma\delta}|$ for any $\gamma$ and $\delta$ with $|\gamma| = |\delta| = k, \ \ 1 \leq k \leq \min(m, n)$. Hence $C_k(eC + fD) = C_k(eC) + C_k(fD)$ for all $1 \leq k \leq \min(m, n)$. We also get that for any $k \geq 1$, $\mathcal{I}_k(eC + fD) \subset \mathcal{I}_k(eC) + \mathcal{I}_k(fD)$. On the other hand $|eC_{\gamma\delta}| = e|eC_{\gamma\delta}| = e(|eC_{\gamma\delta}| + |fD_{\gamma\delta}|) = e|(eC + fD)_{\gamma\delta}|$. Hence $\mathcal{I}_k(eC) \subset \mathcal{I}_k(eC + fD)$. Similarly we also have $\mathcal{I}_k(fD) \subset$

$\mathcal{I}_k(eC+fD)$. Thus $\mathcal{I}_k(eC+fD)=\mathcal{I}_k(eC)+\mathcal{I}_k(fD)$ for all $k \geq 1$. Thus the first part is proved.

Now, (a) follows from the first part.

Let $I(A)$ and $I(B)$ be the Rao-idempotents of $A$ and $B$ respectively. If $r = \rho(A) \geq \rho(B)$ take a $(\gamma,\delta) \in \Delta_{r,m,n}$ such that $|A_{\gamma\delta}| \neq 0$. Then $I(A)|(A+B)_{\gamma\delta}| = I(A)|A_{\gamma\delta}| + I(A)|B_{\gamma\delta}| = I(A)|A_{\gamma\delta}| = |A_{\gamma\delta}| \neq 0$. Hence $|(A+B)_{\gamma\delta}| \neq 0$. Thus $C_r(A+B) \neq 0$. Also $C_{r+1}(A+B) = C_{r+1}(A) + C_{r+1}(B) = 0$. Thus $\rho(A+B) = r$.

Now, if $r=\rho(A)=\rho(B)$ then $(I(A)+I(B)) \in \mathcal{I}_r(A)+\mathcal{I}_r(B) = \mathcal{I}_r(A+B)$ and $(I(A)+I(B))(A+B) = I(A)A+I(B)B = A+B$. Thus $A+B$ is Rao-regular and $I(A+B) = I(A)+I(B)$. Thus (b) is proved.

If $r = \rho(A) > \rho(B)$ and $A$ and $B$ are Rao-regular then $A$ and $B$ are also regular. Let $G$ and $H$ be such that $AGA = A$ and $BHB = B$. Then $(A+B)(I(A)G+I(B)H)(A+B) = A+B$. Thus $A+B$ is regular. Thus (c) is proved.

♠

Generalizing this Proposition to any finite set of Rao-regular matrices with pairwise orthogonal Rao-idempotents gives us

**Proposition 7.13:** Let $D_1, D_2, \cdots, D_\ell$ be $m \times n$ matrices over a commutative ring and let $e_1, e_2, \cdots, e_\ell$ be pairwise orthogonal idempotents.

Then, $C_k(e_1D_1 + e_2D_2 + \cdots + e_\ell D_\ell) = C_k(e_1D_1) + C_k(e_2D_2) + \cdots + C_k(e_\ell D_\ell)$ and $\mathcal{I}_k(e_1D_1 + e_2D_2 + \cdots + e_\ell D_\ell) = \mathcal{I}_k(e_1D_1) + \mathcal{I}_k(e_2D_2) + \cdots + \mathcal{I}_k(e_\ell D_\ell)$ for all $k \geq 1$.

Also if $A_1, A_2, \cdots, A_\ell$ are $m \times n$ Rao-regular matrices over a commutative ring with pairwise orthogonal Rao-idempotents then

(a) $C_k(A_1 + A_2) + \cdots + A_\ell) = C_k(A_1) + C_k(A_2) + \cdots + C_k(A_\ell)$ and $\mathcal{I}_k(A_1 + A_2 + \cdots + A_\ell) = \mathcal{I}_k(A_1) + \mathcal{I}_k(A_2) + \cdots + \mathcal{I}_k(A_\ell)$ for all $k \geq 1$.

(b) If $\rho(A_1) = \rho(A_2) = \cdots = \rho(A_\ell) = r$ then $\rho(A_1 + A_2 + \cdots + A_\ell) = r$ and $A_1 + A_2 + \cdots + A_\ell$ is regular.

(c) $A_1 + A_2 + \cdots + A_\ell$ is regular.

**Proof:** The proof is as in the case of Proposition 7.12.

♠

For a Rao-regular $m \times n$ matrix $A$ even though the Rao-idempotent $I(A) = \sum_{(\alpha,\beta)\in\Delta} |A_{\alpha\beta}|c_{\beta\alpha}$ is unique, the $g$-inverse $G = (g_{ij})$ constructed by the formula $g_{ij} = \sum_{(\alpha,\beta)\in\Delta} \frac{\partial}{\partial a_{ji}}|A_{\alpha\beta}|c_{\beta\alpha}$ (i.e., $G = A_C$ where $C = (c_{\beta\alpha})_{(\alpha,\beta)\in\Delta}$) is an $\binom{n}{r} \times \binom{m}{r}$ matrix, as in section 5.3, will have additional properties depending on the matrix $C$. We shall now look at conditions on $C = (c_{\beta\alpha})_{(\alpha,\beta)\in\Delta}$ so that the constructed $G$ is the Moore-Penrose inverse of $A$, the Group inverse of $A$, etc.

**Theorem 7.14:** Let $R$ be a commutative ring with an involution $a \to \bar{a}$. Let $A$ be an $m \times n$ Rao-regular matrix over $R$ with $\rho(A) = r$ and Rao-idempotent $I(A)$. Let $G$ be an $n \times m$ matrix over $R$. Then in the following, (i) $\Rightarrow$ (ii) $\Rightarrow$ (iii).

(i) $G$ is the Moore-Penrose inverse of $A$.

(ii) $C_r(G)$ is the Moore-Penrose inverse of $C_r(A)$.

(iii) $Tr(C_r(A^*A))Tr(C_r(G^*G)) = I(A)$.

If $Tr(C_r(A^*A))Tr(C_r(G^*G)) = I(A)$ then $A^+ = A_{Tr(C_r(G^*G))C_r(A^*)}$.

**Proof:** (i) $\Rightarrow$ (ii) is clear.

Since $\rho(C_r(A)) = 1$, by the last part of Theorem 7.3 we get that $Tr(C_r(A^*A))Tr(C_r(G^*G))C_r(A) = C_r(A)$. Since $Tr(C_r(A^*A))Tr(C_r(G^*G)) = Tr(C_r(A^*)C_r(A))$, and $Tr(C_r(G^*G)) \in \mathcal{I}_r(A)$ we have from Proposition 7.11 that $I(A) = I(C_r(A)) = Tr(C_r(A^*A))Tr(C_r(G^*G))$. Thus (ii) $\Rightarrow$ (iii) is proved.

Now let $H = A_{Tr(C_r(G^*G))C_r(A^*)}$.

Suppose also that $Tr(C_r(G^*G))Tr(C_r(A^*A)) = I(A)$. From Theorem 5.5 we have $AHA = Tr(C_r(A)Tr(C_r(G^*G))C_r(A^*))A = I(A)A = A$. Since $\rho(Tr(C_r(G^*G))C_r(A^*)) = 1$ again from Theorem 5.5 we have $HAH = H$. Since $(C_r(A)Tr(C_r(G^*G))C_r(A^*))^* = C_r(A)Tr(C_r(G^*G))C_r(A^*)$ we have from Theorem 5.5 that $(AH)^* = AH$. Also $(HA)^* = HA$ is clear from Theorem 5.5. Thus $H$ is the Moore-Penrose inverse of $A$.

♠

Let us now give simple necessary and sufficient conditions for a Rao-regular matrix to have Moore-Penrose inverse.

**Proposition 7.15:** Let $R$ be a commutative ring with an involution $a \to \bar{a}$. Let $A$ be an $m \times n$ Rao-regular matrix over $R$ with $\rho(A) = r$ and Rao-idempotent $I(A)$. Then the following are equivalent.

(i) A has Moore-Penrose inverse.

(ii) $\overline{I(A)} = I(A)$ and $Tr(C_r(A^*A))|I(A)$.

(iii) $\overline{I(A)} = I(A)$ and $Tr(C_r(A^*A))$ has a $g$-inverse $v$ in $R$ such that $Tr(C_r(A^*A))v = I(A)$.

(iv) $Tr(C_r(A^*A))$ has a $g$-inverse $w$ in $R$ such that $\bar{w} = w$ and $Tr(C_r(A^*A))w = I(A)$.

(v) $(Tr(C_r(A^*A)))^+$ exists in $R$ and $Tr(C_r(A^*A))(Tr(C_r(A^*A)))^+ = I(A)$ (i.e., $Tr(C_r(A^*A))(Tr(C_r(A^*A)))^+A = A$).

In case $w$ is an element of $R$ such that $Tr(C_r(A^*A))w = I(A)$ and $\bar{w} = w$ then $A^+ = A_{wC_r(A^*)}$.

**Proof:** If $G$ is the Moore-Penrose inverse of $A$ then from Theorem 7.14 we have that $Tr(C_r(A^*A))Tr(C_r(G^*G)) = I(A)$. Hence $\overline{I(A)} = I(A)$ and $Tr(C_r(A^*A))|I(A)$. Thus (i) $\Rightarrow$ (ii).

Let $v$ be an element of $R$ such that $Tr(C_r(A^*A))v = I(A)$. Since $Tr(C_r(A^*A)) \in \mathcal{I}_r(A)$ and $I(A)$ is the identity element of $\mathcal{I}_r(A)$, we have $I(A)Tr(C_r(A^*A)) = Tr(C_r(A^*A))$. Hence $Tr(C_r(A^*A))vTr(C_r(A^*A) = Tr(C_r(A^*A))$. Thus $v$ is a $g$-inverse of $Tr(C_r(A^*A))$ such that $Tr(C_r(A^*A))v = I(A)$. Thus (iii) is proved from (ii).

If $v$ is a $g$-inverse of $Tr(C_r(A^*A))$ such that $Tr(C_r(A^*A))v = I(A)$ and $\overline{I(A)} = I(A)$ then $w = vTr(C_r(A^*A))\bar{v}$ is a $g$-inverse of $Tr(C_r(A^*A))$ such that $\bar{w} = w$ and $Tr(C_r(A^*A))w = I(A)$. Thus (iv) follows from (iii).

If $w$ is a $g$-inverse of $Tr(C_r(A^*A))$ such that $Tr(C_r(A^*A))w = I(A)$ and $\bar{w} = w$ then the element $wTr(C_r(A^*A))w$ is a reflexive $g$-inverse of $Tr(C_r(A^*A))$ and it also satisfies $(Tr(C_r(A^*A))wTr(C_r(A^*A))w)^* = Tr(C_r(A^*A))wTr(C_r(A^*A))w$. Thus $vTr(C_r(A^*A))w = (Tr(C_r(A^*A)))^+$. Thus (v) follows from (iv).

If $w$ is any element of $R$ such that $\bar{w} = w$ and $Tr(C_r(A^*A))w = I(A)$ then as shown in the proof of Theorem 7.14 it follows that $A^+ = A_{wC_r(A^*)}$. Thus (v) $\Rightarrow$ (i) is proved and also the last statement of the Theorem.

♠

For a Rao-regular matrix to admit a group inverse we have,

**Theorem 7.16:** Let $R$ be a commutative ring. Let $A$ be an $m \times m$ Rao-regular matrix over $R$ with $\rho(A) = r$ and Rao-idempotent $I(A)$. Let $G$ be an $m \times m$ matrix over $R$. Then in the following, (i) $\Rightarrow$ (ii) $\Rightarrow$ (iii)$\Rightarrow$(iv).

(i) $G$ is the group inverse of $A$.

(ii) $C_r(G)$ is the group-inverse of $C_r(A)$.

(iii) $Tr(C_r(G))Tr(C_r(A)) = I(A)$.

(iv) $(Tr(C_r(G)))^2(Tr(C_r(A)))^2 = I(A)$.

Further if $(Tr(C_r(G)))^2(Tr(C_r(A)))^2 = I(A)$ (or $Tr(C_r(G))Tr(C_r(A)) = I(A)$) then $A^\# = A_{(Tr(C_r(G)))^2C_r(A)}$.

**Proof:** (i) $\Rightarrow$ (ii) is clear.

Since $\rho(C_r(A)) = 1$, by Theorem 7.5 we get that $Tr(C_r(G))C_r(A) = C_r(G)C_r(A)$. Hence $Tr(C_r(G))Tr(C_r(A)) = Tr(C_r(G)C_r(A)) = I(A)$. Thus (ii) $\Rightarrow$ (iii) is proved.

(iii) $\Rightarrow$ (iv) is clear because $I(A)$ is idempotent.

Now, let $w = (Tr(C_r(G)))^2$ and let $H = A_{wC_r(A)}$. Suppose also that $(Tr(C_r(G)))^2(Tr(C_r(A)))^2 = w(Tr(C_r(A)))^2 = I(A)$. From Theorem 5.5 we have $AHA = (Tr(C_r(G)))^2(Tr(C_r(A)))^2A = I(A)A = A$. Since $\rho(wC_r(A)) = 1$, again from Theorem 5.5 we have $HAH = H$. That $AH = HA$ is clear from Theorem 5.5 since $wC_r(A)$ commutes with $C_r(A)$. Thus $H$ is the group inverse of $A$.

♠

We shall now give simple necessary and sufficient conditions for a Rao-regular matrix to have group inverse.

**Proposition 7.17:** Let $R$ be a commutative ring. Let $A$ be an $m \times m$ Rao-regular matrix over $R$ with $\rho(A) = r$ and Rao-idempotent $I(A)$. Then the following are equivalent.

(i) $A$ has group inverse.

(ii) $Tr(C_r(A))|I(A)$.

(iii) $(Tr(C_r(A)))^2|I(A)$.

(iv) $Tr(C_r(A))$ is a regular element of $R$ and $Tr(C_r(A))(Tr(C_r(A)))^- = I(A)$ for some $g$-inverse $(Tr(C_r(A)))^-$ of $Tr(C_r(A))$.

In case $Tr(C_r(A))|I(A)$, if $v$ is an element of $R$ such that $Tr(C_r(A))v = I(A)$ then $A^\# = A_{v^2C_r(A)}$.

In case $(Tr(C_r(A)))^2|I(A)$, if $w$ is an element of $R$ such that $(Tr(C_r(A)))^2w = I(A)$ then $A^\# = A_{wC_r(A)}$.

**Proof:** If $G$ is the group inverse of $A$ we have from Theorem 7.16 that $Tr(C_r(A))Tr(C_r(G)) = I(A)$. Thus $Tr(C_r(A))|I(A)$. Thus (i) $\Rightarrow$ (ii). (ii) $\Rightarrow$ (iii) follows because $I(A)$ is idempotent.

In case $(Tr(C_r(A)))^2w = I(A)$ then, following the last part of the proof of Theorem 7.16 we get that $A^\# = A_{wC_r(A)}$. Thus (iii) $\Rightarrow$ (i).

(ii) $\Rightarrow$ (iv) follows because $I(A)C_r(A) = C_r(A)$. (iv) $\Rightarrow$ (ii) is clear.

♠

As a final result of this section we shall show that the group inverse of a Rao-regular matrix $A$ is a polynomial in $A$.

**Proposition 7.18:** If $A$ is an $m \times m$ Rao-regular matrix over a commutative ring $R$ with $\rho(A) = r$ and with a group inverse $A^\#$ then $A^\#$ is a polynomial in $A$.

**Proof:** Let $e = I(A)$ be the Rao-idempotent of $A$. Then $A$ is a matrix over the ring $eR$ and $e$ is the identity of $eR$. By Proposition 7.17 (ii) we get that $Tr(C_r(A))$ is a unit of $eR$. By the later part of Theorem 6.21 it follows that the group inverse of $A$ is a polynomial in $A$.

♠

Thus we have completed a study of Rao-regular matrices.

## 7.4 Regular matrices over commutative rings

We shall now characterize regular matrices over commutative rings. The characterization will be in terms of Rao-regular matrices.

In Proposition 7.13 we have seen that every sum of Rao-regular matrices with orthogonal Rao-idempotents is regular. We shall show in the decomposition Theorem below that every regular matrix can be written as a sum of Rao-regular matrices with orthogonal Rao-idempotents. The initial step was already observed in Proposition 7.10, viz., every regular matrix has a Rao-regular component.

**Theorem 7.19 (Canonical Decomposition Theorem of Prasad):** Let $A$ be an $m \times n$ matrix over a commutative ring $R$ with $\rho(A) = r > 0$. Then the following are equivalent.

(i) $A$ is regular;

(ii) There exists a $k \geq 1$ and nonzero idempotents $e_1, e_2, \cdots, e_k$ such that

(a) $A = e_1A + e_2A + \cdots + e_kA$;

(b) $e_ie_j = 0$ for all $i \neq j$;

(c) $e_1A, e_2A, \cdots, e_kA$ are all Rao-regular with $I(e_iA) = e_i$ for all $1 \leq i \leq k$; and

(d) $r = \rho(e_1A) > \rho(e_2A) > \cdots > \rho(e_kA) > 0$.

(iii) There exists a $k \geq 1$ and matrices $A_1, A_2, \cdots, A_k$ such that

(a) $A = A_1 + A_2 + \cdots + A_k$;

(b) $\{A_1, A_2, \cdots, A_k\}$ are Rao-regular with orthogonal Rao-idempotents; and

(c) $r = \rho(A_1) > \rho(A_2) > \cdots > \rho(A_k) > 0$.

Also, $k$ and $e_1, e_2, \cdots, e_k$ of (ii) when they exist are unique. $k$ and $A_1$, $A_2, \cdots, A_k$ of (iii) when they exist are also unique. The $k$ of (ii) and $k$ of (iii) are equal. If (iii) holds $e_i = I(A_i)$ for $1 \leq i \leq k$ serve the purpose of (ii). If (ii) holds $A_i = e_iA$ for $1 \leq i \leq k$ serve the purpose of (iii).

**Proof:** (i) $\Rightarrow$ (ii). We shall prove this by induction on $\rho(A)$. If $\rho(A) = 1$ then $A$ is Rao-regular by Theorem 7.9 (d). If we take $e_1 = I(A)$ and $k = 1$ we get that $A = e_1A$ and (ii) is proved.

If $r_1 = r = \rho(A) > 1$, since $A$ is regular, by Proposition 7.7 (a) there is an $e_1 \in \mathcal{I}_r(A)$ such that $e_1$ is the identity of $\mathcal{I}_r(A)$. By Proposition 7.10 this idempotent $e_1$ has the properties that $\rho(e_1A) = r_1$, $e_1A$ is Rao-regular, $I(e_1A) = e_1, \rho((1 - e_1)A) < r$ and $(1 - e_1)A$ is regular.

Let $\rho((1 - e_1)A) = r_2$. If $r_2 = 0$ then $A = e_1A$ and $A$ is Rao-regular. Hence $k = 1$ and $e_1$ serve the purpose of (ii). If $r_2 > 0$, by the induction hypothesis there exist nonzero idempotents $e_2, e_3, \cdots, e_k$ such that $(1 - e_1)A = e_2(1 - e_1)A + e_3(1 - e_1)A + \cdots + e_k(1 - e_1)A$, $e_ie_j = 0$ for all $i \neq j$, $i, j \geq 2$, $e_2(1-e_1)A$, $e_3(1-e_1)A, \cdots, e_k(1-e_1)A$ are all Rao-regular with $I(e_i(1-e_1)A) = e_i$ for all $2 \leq i \leq k$ and $r_2 = \rho(e_2(1-e_1)A) > \rho(e_3(1-e_1)A) > \cdots > \rho(e_k(1-e_1)A) > 0$.

But, for $2 \leq i \leq k$, since $e_i = I(e_i(1-e_1)A)$, there exists an $x \in \mathcal{I}_{r_i}(A)$ where $r_i = \rho(e_i(1-e_1)A)$ such that $e_i = e_i(1-e_1)x$. Hence $e_i(1-e_1) = e_i(1-e_1)x = e_i$ for $2 \leq i \leq k$.

Hence for $e_1, e_2, \cdots, e_k$, we have $A = e_1A + e_2A + \cdots + e_kA$; $e_ie_j = 0$ for $1 \leq i, j \leq k$; $e_1A, e_2A, \cdots, e_kA$ are all Rao-regular with $I(e_iA) = e_i$ for all $1 \leq i \leq k$ and $r_1 = r = \rho(e_1A) > \rho(e_2A) > \cdots > \rho(e_kA)$. Thus (ii) is proved from (i).

(ii) $\Rightarrow$ (iii) is clear by taking $A_i = e_iA$ for $1 \leq i \leq k$.

(iii) $\Rightarrow$ (i) follows from Proposition 7.13.

To prove the uniqueness of $k, e_1, e_2, \cdots, e_k$ satisfying (ii) observe that since $e_1A, e_2A, \cdots, e_kA$ are all Rao-regular matrices with orthogonal Rao-idempotents, by Proposition 7.13 we have that $\mathcal{I}_\ell(A) = \mathcal{I}_\ell(e_1A+\cdots+e_kA) = \mathcal{I}_\ell(e_1A)+\mathcal{I}_\ell(e_2A)+\cdots+\mathcal{I}_\ell(e_kA)$ for all $\ell \geq 0$. If we take $\ell = r = \rho(A)$, since $r = \rho(e_1A) > \rho(e_2A) > \cdots > \rho(e_kA) > 0$, we have $\mathcal{I}_r(A) = \mathcal{I}_r(e_1A)$.

But $e_1$ being the Rao-idempotent of $e_1A$, $e_1$ is the identity of $\mathcal{I}_r(A)$. This shows that $e_1$ is unique. Starting with $A - e_1A = e_2A + e_3A + \cdots + e_kA$ it follows that $e_2$ is unique. Continuing this argument we get that $e_1, e_2, \cdots, e_k$ and $k$ are unique.

To prove the uniqueness of $k$ and $A = A_1 + A_2 + \cdots + A_k$ in (iii), observe that, if we call $I(A_i) = e_i$ for $1 \leq i \leq k$ then $e_iA_j = e_ie_jA_j = 0$ for $i \neq j$. Hence we have $e_iA = e_iA_i = A_i$. Hence $A = e_1A + e_2A + \cdots + e_kA$ is the decomposition of (ii). But since the decomposition of (ii) is already shown to be unique we have the uniqueness of $k$ and $A_1, A_2, \cdots, A_k$ in (iii).

The rest of the parts of the Theorem are clear.

♠

We shall call the decomposition of $A$ as in (ii) the canonical decomposition of $A$. Thus we have shown a canonical decomposition of any regular matrix over a commutative ring. Given a canonical decomposition of a regular matrix let us find the canonical decompositions of various other related regular matrices.

**Proposition 7.20:** Let $A$ be a regular matrix over a commutative ring $R$ with $\rho(A) > 0$. Let $A = e_1A+e_2A+\cdots+e_kA$ be the canonical decomposition of $A$. Let $G$ be a $g$-inverse of $A$. Then $AG = e_1AG + e_2AG + \cdots + e_kAG$ and $GA = e_1GA + e_2GA + \cdots + e_kGA$ are the canonical decompositions of $AG$ and $GA$ respectively. If $G$ is a reflexive $g$-inverse of $A$ then $G = e_1G + e_2G + \cdots + e_kG$ is the canonical decomposition of $G$.

**Proof:** Let $A = e_1A + e_2A + \cdots + e_kA$ be the canonical decomposition of $A$. Let $G$ be a $g$-inverse of $A$. Then for any $i$ with $1 \leq i \leq k$, since $AGA = A$, we have $e_iAGe_iA = e_iA$. Thus $G$ is a $g$-inverse of $e_iA$ too. From (b) of Proposition 7.11 we have that $e_iAG$ and $e_iGA$ are also Rao-regular and $e_i = I(e_iA) = I(e_iAG) = I(e_iGA)$. Also $\rho(e_iAG) = \rho(e_iA)$. Hence, $AG = e_1AG + e_2AG + \cdots + e_kAG$ and $GA = e_1GA + e_2GA + \cdots + e_kGA$ are canonical decompositions of $AG$ and $GA$ respectively.

If $G$ is a reflexive $g$-inverse of $A$ then for any $i$ with $1 \leq i \leq k$, since $e_i$ is idempotent, $e_iG$ is a reflexive $g$-inverse of $e_iA$ and $\rho(e_iA) = \rho(e_iG)$. From part (c) of Proposition 7.11 it follows that $e_iG$ is Rao-regular and $e_i = I(e_iA) = I(e_iG)$. Also $G = GAG = e_1GAG + e_2GAG + \cdots + e_kGAG = e_1Ge_1Ae_1G + e_2Ge_2Ae_2G + \cdots + e_kGe_kAe_kG = e_1G + e_2G + \cdots + e_kG$. Thus $G = e_1G + e_2G + \cdots + e_kG$ is the canonical decomposition of $G$.

♠

If $A = e_1A + e_2A + \cdots + e_kA$ is the canonical decomposition of $A$ and if for every $1 \leq i \leq k, G_i$ is a $g$-inverse of $e_iA$ then $G_1 + G_2 + \cdots + G_k$ is a $g$-inverse of $A$. Since the construction of $g$-inverses for Rao-regular matrices is similar to the construction of $g$-inverses in the case of integral domains, it is possible to give the complete construction of all $g$-inverses of a regular matrix. We shall do this in Theorem 7.31.

In Theorem 7.19 we have seen that every regular matrix can be decomposed in a canonical way into Rao-regular matrices. In a similar way it is also possible to write every $m \times n$ matrix over a commutative ring as a sum of some Rao-regular matrices and a non-Rao-regular matrix. We shall obtain this now.

**Theorem 7.21 (Robinson's decomposition Theorem):** Let $A$ be an $m \times n$ matrix over a commutative ring $R$ with $\rho(A) = r$. Then there exists a unique integer $t \geq 1$ and a unique list $(e_1, e_2, \cdots, e_t)$ of pairwise orthogonal idempotents of $R$ such that, if $r_i = \rho(e_iA)$ then

(i) $e_1 + e_2 + \cdots + e_t = 1$;

(ii) $r = r_1 > r_2 > \cdots > r_\ell \geq 0$;

(iii) for $1 \leq j < t, e_jA$ is Rao-regular with Rao-idempotent $I(e_jA) = e_j$; and

(iv) either $e_tA = 0$ or $\mathcal{I}_{r_t}(e_tA)$ does not posses an identity element.

**Proof:** We shall prove this by induction on $r = \rho(A)$. If $\rho(A) = 0$, i.e., $A = 0$ then $t = 1$ and $e_1 = 1$ satisfy (i)–(iv) of the Theorem.

Now, let $\rho(A) = r > 0$. Let us assume that there exist $t$ and $(e_1, e_2, \cdots, e_t)$ satisfying (i)–(iv) for every matrix of rank $< r$.

If $\mathcal{I}_r(A)$ does not possess an identity element then $t = 1$ and $e_1 = 1$ satisfy (i)–(iv). If $\mathcal{I}_r(A)$ does possess an identity element, say $e$, then by

Proposition 7.10, $eA$ is Rao-regular, $\rho(eA) = r$ and $\rho((1-e)A) < r$. By the induction hypothesis, for the matrix $(1-e)A$, there exist $t \geq 1$ and a pairwise orthogonal list of idempotents $(f_1, f_2, \cdots, f_t)$ satisfying (i)–(iv) of the Theorem. For $1 \leq j \leq t$ let $k_j = \rho(f_j(1-e)A)$. For $1 \leq j < t$, $f_j(1-e)A$ is Rao-regular and $I(f_j(1-e)A) = f_j$. But, since $f_j \in \mathcal{I}_{k_j}(f_j(1-e)A)$ we get that $f_j(1-e) = f_j$ for $1 \leq j < t$. This in turn would give us that $f_t(1-e) = f_t$.

Thus $1 = e+(1-e) = e+(1-e)(f_1+f_2+\cdots+f_t) = e+f_1+f_2+\cdots+f_t$ and $(e, f_1, f_2, \cdots, f_t)$ is the list of idempotents for $A$ – because – they are clearly pairwise orthogonal, $r = \rho(eA) > \rho(f_1A) > \rho(f_2A) > \cdots > \rho(f_tA)$ and (iii) and (iv) are also satisfied. Thus $t+1$ and $(e, f_1, f_2, \cdots, f_t)$ serve the purpose for $A$.

We shall now see the uniqueness. If $\rho(A) = 0$ and $(e_1, e_2, \cdots, e_t)$ is a list of idempotents satisfying (i)–(iv) for $A$ then unless $t = 1, 0 = \rho(e_1A) > \rho(e_2(A)) \geq 0$ will be a violation. Once we see that $t = 1$ clearly $e_1 = 1$. Thus the uniqueness of the list of idempotents in the case $A = 0$ is proved.

If $\rho(A) = r > 0$ and $e_1 = 1$ is the list of idempotents for $A$ then $\mathcal{I}_r(A)$ does not possess an identity element. If $(f_1, f_2, \cdots, f_s)$ is another list of idempotents for $A$ satisfying (i)–(iv), by Proposition 7.12 we have that $C_r(A) = C_r(f_1A)$ and $\mathcal{I}_r(A) = \mathcal{I}_r(f_1A)$. If $s \geq 2$ we get that $f_1$ is the identity element of $\mathcal{I}_r(A)$. Hence $s = 1$ and $f_1 = 1$. Thus the uniqueness of the list of idempotents in the case of $\rho(A) = r > 0$ and $\mathcal{I}_r(A)$ not possessing identity element is established.

Now suppose that $\rho(A) = r > 0$ is such that $(e_1, e_2, \cdots, e_t), (f_1, f_2, \cdots, f_s)$ both are lists of idempotents of $A$ satisfying (i)–(iv) and $2 \leq t \leq s$. As we have seen before, it follows that $e_1 = f_1, e_2 = f_2, \cdots, e_{t-1} = f_{t-1}$. Now, if $e_tA = 0$ then $e_tA = A - e_1A - e_2A - \cdots - e_{t-1}A = A - f_1A - f_2A - \cdots - f_{t-1}A = f_tA + \cdots + f_sA$. Since $\rho(f_tA) > \rho(f_{t+1}A) \geq 0$ unless $s = t$, we have that $s = t$ and $f_tA = 0$. Hence $f_t = e_t$. If $e_tA \neq 0$ and $\mathcal{I}_{r_t}(e_tA)$ does not possess an identity then by the previous argument we have that $s = t$ and $f_t = e_t$. Thus the uniqueness is proved.

♠

Thus every matrix $A$ has a decomposition $A = e_1A + \cdots + e_tA$. Following [88] we shall call $t$ the *Rao-index*, $(e_1, e_2, \cdots, e_t)$ the *Rao-list of idempotents* and $r_1, r_2, \cdots, r_t$ the *Rao-list of ranks* of $A$. Now,

**Proposition 7.22:** Let $A$ be an $m \times n$ matrix over a commutative ring $R$ with Rao-index $t$ and Rao-list of idempotents $(e_1, e_2, \cdots, e_t)$. Then $A$ is regular if and only if $e_tA = 0$. $A$ is Rao-regular if and only if $t = 2$ and $e_2A = 0$.

**Proof:** This is clear.

♠

It is also clear that if $A$ is regular with Rao-list of idempotents $(e_1, e_2, \cdots, e_t)$ then $A = e_1A + e_2A + \cdots + e_{t-1}A$ is the canonical decomposition of $A$. In the later sections we shall make use of the decomposition Theorem of this section. As an application of the previous Proposition, we shall now derive Von Neumann's result of section 3.3 for commutative rings. Recall that a ring $R$ is regular if every element of $R$ is regular. We need an elementary result.

**Proposition 7.23:** Every ideal generated by a finite number of regular elements in a commutative ring has the identity element. In particular, Every finitely generated nonzero ideal in a commutative regular ring has the identity element.

**Proof:** Let $\mathcal{I}$ be an ideal generated by regular elements $\{z_1, z_2, \cdots, z_k\}$. If for $1 \leq i \leq k, z_i^-$ is a $g$-inverse of $z_i$ then $\mathcal{I}$ is generated by $\{z_1z_1^-, z_2z_2^-, \cdots, z_kz_k^-\}$. But for $1 \leq i \leq k, z_iz_i^-$ is idempotent. Let $e_i = z_iz_i^-$ for $1 \leq i \leq k$. Then $\mathcal{I}$ is generated by $\{e_1, (1-e_1)e_2, (1-e_2)(1-e_1)e_3, \cdots, (1-e_{k-1})(1-e_{k-2})\cdots(1-e_1)e_k\}$. For $1 \leq i \leq k$ if we write $f_i = (1-e_{i-1})(1-e_{i-2})\cdots(1-e_1)e_i$ with the understanding that $f_1 = e_1$ then $f_if_j = 0$ for $i \neq j$. Hence $f_1 + f_2 + \cdots + f_k$ is the identity of the ideal $\mathcal{I}$.

♠

Now,

**Theorem 7.24:** If $R$ is a commutative regular ring then every $m \times n$ matrix over $R$ is regular.

**Proof:** Let $A$ be an $m \times n$ non zero matrix over $R$. From Robinson's decomposition Theorem (Theorem 7.21) $A$ can be written as $e_1A + e_2A + \cdots + e_tA$ where $r_t = \rho(e_tA) \geq 0$ and either $e_tA = 0$ or $\mathcal{I}_{r_t}(e_tA)$ does not possess an identity element. If $r_t > 0$ then $\mathcal{I}_{r_t}(e_tA)$ is a finitely generated ideal over a commutative regular ring and by Proposition 7.23 this ideal possesses the identity element. Hence $e_tA = 0$. By Proposition 7.22 this means that $A$ is regular.

♠

As another application of Proposition 7.22, let us give a sufficient condition for a matrix A to be regular. Incidentally, Von Neumann's result can be derived from the next result too.

**Proposition 7.25:** Let $R$ be a commutative ring. Let $A$ be an $m \times n$ matrix over $R$ with $\rho(A) = r$. If each $|A_{\alpha\beta}| : |\alpha| \leq r, |\beta| \leq r, |\alpha| = |\beta|$ is regular then $A$ is regular.

**Proof:** The hypothesis tells us that for any fixed $s$ such that $1 \leq s \leq r$, $\mathcal{I}_s(A)$ is an ideal generated by regular elements. Hence from Proposition 7.23 we get that $\mathcal{I}_s(A)$ has the identity.

Now, if we apply Robinson's Decomposition Theorem to $A$ and write $A = e_1A + e_2A + \cdots + e_tA$ then $e_tA = 0$, because, $\mathcal{I}_{r_t}(e_tA) = e_t\mathcal{I}_{r_t}(A)$ has the identity by Proposition 7.23.

Thus $A$ is regular by Proposition 7.22.

♠

## 7.5 All generalized inverses

In Chapter 6 we have seen that for an $m \times n$ matrix A over an integral domain, if $S$ is an $\binom{n}{r} \times \binom{m}{r}$ matrix such that $Tr(SC_r(A)) = 1$ then $A_S$ is a $g$-inverse of $A$. The question arises as to whether every $g$-inverse of $A$ is an $A_S$ for some $S$ with $Tr(SC_r(A)) = 1$. We shall give an affirmative answer to this after considering related questions for matrices over commutative rings.

Over an integral domain we have seen in Theorem 6.2 that if $G$ is a reflexive $g$-inverse of $A$ then $G = A_{C_r(G)}$. This result is not true for matrices over general commutative rings (see section 8.3). We shall now extend this result to Rao-regular matrices over commutative rings.

**Theorem 7.26:** Let $A$ be an $m \times n$ Rao-regular matrix over a commutative ring with $\rho(A) = r$. If $G$ is a reflexive $g$-inverse of $A$ then $G = A_{C_r(G)}$.

**Proof:** Let $I(A)$ be the Rao-idempotent of $A$. Since $G$ is a $g$-inverse of $A$, by Proposition 7.11 $AG$ is also Rao-regular and $I(AG) = I(A) = Tr(C_r(AG))$.

Also, since an idempotent matrix is its own group inverse, we have that $AG$ is the group inverse of $AG$. Hence by Theorem 7.17, $AG = (AG)_{wC_r(AG)}$ where $w$ is any element such that $(Tr(C_r(AG)))^2w = I(A) = Tr(C_r(AG))$. Since $w$ can be taken to be 1, the unit element of $R$, we have $AG = (AG)_{C_r(AG)}$, i.e., if $AG = E = (e_{ij})$ then

$$e_{kj} = \sum_{\alpha,\beta:(\alpha,\beta)\in\Delta} |(AG)_{\beta\alpha}| \frac{\partial|(AG)_{\alpha\beta}|}{\partial e_{jk}}.$$

If $G = (g_{ij})$ and $G$ is a reflexive $g$-inverse of $A$ we have,

$$g_{ij} = \sum_k g_{ik}e_{kj} = \sum_k g_{ik} \sum_{\alpha,\beta:(\alpha,\beta)\in\Delta} |(AG)_{\beta\alpha}| \frac{\partial|(AG)_{\alpha\beta}|}{\partial e_{jk}}$$

$$= \sum_{\alpha,\beta:(\alpha,\beta)\in\Delta} |(AG)_{\beta\alpha}| \sum_k g_{ik} \frac{\partial|(AG)_{\alpha\beta}|}{\partial e_{ik}}.$$

Note that for all $\alpha$ for which $j \in \alpha$ and all $\beta$, $\sum_k g_{ik} \frac{\partial}{\partial e_{ik}} |(AG)_{\alpha\beta}| = |K_{\alpha\beta}|$ where $K$ is the matrix obtained from $AG$ by replacing the $j$th row of $AG$ with the $i$th row of $G$. If we call as $B$ the matrix obtained from $A$ by replacing the $j$th row of $A$ with the row $(0, 0, \cdots, 0, 1, 0, \cdots, 0)$ where the $i$th entry is 1 and all other entries are zero, then we have $K = BG$. Hence

$$|K_{\alpha\beta}| = \sum_{\gamma \in Q_{r,n}} |B_{\alpha\gamma}||G_{\gamma\beta}| = \sum_{\gamma \in Q_{r,n}} \frac{\partial |A_{\alpha\gamma}|}{\partial a_{ji}} |G_{\gamma\beta}|.$$

Hence

$$\begin{aligned} g_{ij} &= \sum_{\alpha,\beta:(\alpha,\beta)\in\Delta} |(AG)_{\beta\alpha}| \sum_{\gamma\in Q_{r,n}} \frac{\partial |A_{\alpha\gamma}|}{\partial a_{ji}} |G_{\gamma\beta}| \\ &= \sum_{\alpha,\beta,\gamma:j\in\alpha} |G_{\gamma\beta}||AG|_{\beta\alpha} \frac{\partial |A_{\alpha\gamma}|}{\partial a_{ji}} \\ &= \sum_{\alpha,\gamma:j\in\alpha} |G_{\gamma\alpha}| \frac{\partial |A_{\alpha\gamma}|}{\partial a_{ji}}. \end{aligned}$$

Hence $G = A_{C_r(G)}$.

♠

In our next Theorem we show that every $g$-inverse of an $m \times n$ Rao-regular matrix $A$ is of the form $A_S$ for some $\binom{n}{r} \times \binom{m}{r}$ matrix $S$. We need a few technical results which are interesting by themselves too.

**Proposition 7.27:** Let $A$ be a nonzero $m \times n$ Rao-regular matrix over a commutative ring with $\rho(A) = r$ and Rao-idempotent $I(A)(= e$, say). Let $B$ be any matrix of any order. Then every matrix of the form $eB$ can be written as $\sum_{\theta\in\Theta} X_\theta A Y_\theta$ for some matrices $\{X_\theta, Y_\theta\}_{\theta\in\Theta}$ of appropriate orders.

**Proof:** Firstly, observe that since $e \in \mathcal{I}_r(A) \subset \mathcal{I}_1(A)$ there exists elements $\{c_{ij}\}_{1\le i\le m, 1\le j\le n}$ such that $\sum_{\substack{1\le i\le m \\ 1\le j\le n}} c_{ij}a_{ij} = e$.

If $I$ is the $k \times k$ identity matrix for some $k$, we shall show that $eI$ can be written as $\sum_{\theta\in\Theta} X_\theta A Y_\theta$ for some $\{X_\theta, Y_\theta\}_{\theta\in\Theta}$. Then, clearly $eB = eIB = \sum_{\theta\in\Theta} X_\theta A Y_\theta B$ and hence the Proposition would follow for every $eB$.

For $s, t$, let $\Lambda_{st}$ be the $k \times k$ matrix with 1 in the $(s, t)$ position and zero elsewhere. Again, for $s, t$, let $\Gamma_{st}$ be the $k \times m$ matrix with 1 in the $(s, t)$ position and zero elsewhere. Similarly, for $s, t$, let $\Delta_{st}$ be the $n \times k$ matrix with 1 in the $(s, t)$ position and zero elsewhere.

Now, a little bit of calculation gives us that $e\Lambda_{st} = \sum_{\substack{1\le i\le m \\ 1\le j\le n}} c_{ij}\Gamma_{si}A\Delta_{jt}$. Hence,

$$eI = e\Lambda_{11} + e\Lambda_{22} + \cdots + e\Lambda_{kk} = \sum_{1\le \ell\le k}\ \sum_{\substack{1\le i\le m \\ 1\le j\le n}} c_{ij}\Gamma_{\ell i}A\Delta_{j\ell}.$$

Thus $eI$ is of the form $\sum_{\theta\in\Theta} X_\theta AY_\theta$ for some matrices $\{X_\theta, Y_\theta\}_{\theta\in\Theta}$.

♠

Our next result, though still technical is more interesting.

**Proposition 7.28:** Let $A$ be an $m \times n$ Rao-regular matrix over a commutative ring with $\rho(A) = r$ and Rao-idempotent $I(A)$ $(= e$, say). Then every $g$-inverse of $A$ of the form $eC$ where $C$ is any $n \times m$ can be written as $\sum_{j=1}^{k} \varepsilon_i H_i$ where $\{H_i, 1 \le i \le k\}$ are all reflexive $g$-inverses of $A$ and for $1 \le i \le k, \varepsilon_i = \pm 1$.

**Proof:** Let $eC = G$ and $H = GAG$. Then $H$ is a reflexive $g$-inverse of $A$. As in Theorem 4.19, let us define $H(X,Y)$ for any two matrices $X$ and $Y$ of appropriate orders by

$$H(X,Y) = H + (I - HA)XAH + HAY(I - AH) + (I - HA)XAY(I - AH).$$

Then $H(X,Y)$ is a reflexive $g$-inverse of $A$ for every $X$ and $Y$. Also, note that

$$(I - HA)XAY(I - AH) = H(0,0) - H(X,0) - H(0,Y) + H(X,Y).$$

Observe that $(I - HA)(G - H)(I - AH) = G - H$. Hence $G = (I - HA)(G - H)(I - AH) + H$. But $G - H$ being a matrix of the form $eB$ can be written, by Proposition 7.27, as $\sum_{\theta\in\Theta} X_\theta AY_\theta$ for some matrices $\{X_\theta, Y_\theta\}_{\theta\in\Theta}$ of appropriate orders. Hence

$$\begin{aligned} G &= \sum_{\theta\in\Theta}(I - HA)X_\theta AY_\theta(I - AH) + H \\ &= \sum_{\theta\in\Theta}(H(0,0) - H(X_\theta,0) - H(0,Y_\theta) + H(X_\theta,Y_\theta)) + H. \end{aligned}$$

Thus $G = \sum_{i=1}^{k} \varepsilon_i H_i$ for some $k$ where $H_i, 1 \le i \le k$ are reflexive $g$-inverses of $A$ and for $1 \le i \le k, \varepsilon_i = \pm 1$.

♠

Let us now show that for a Rao-regular matrix $A$, every $g$-inverse of $A$ of the form $eC$ is $A_S$ for some $\binom{n}{r} \times \binom{m}{r}$ matrix $S$.

**Theorem 7.29:** Let $A$ be a Rao-regular matrix over a commutative ring with Rao-identity $I(A) = e$. If $eC$ is a $g$-inverse of $A$ then there is an $\binom{n}{r} \times \binom{m}{r}$ matrix $S$ such that $eC = A_S$ and for this $S$, $Tr(SC_r(A)) = e$.

**Proof:** We have seen in the previous Proposition that $eC = \sum_{i=1}^{k} \varepsilon_i H_i$ for some $k$, for some reflexive $g$-inverses $H_i, 1 \leq i \leq k$ of $A$ and $\varepsilon_i = \pm 1$ for $1 \leq i \leq k$. Also, by Proposition 7.26, $H_i = A_{C_r(H_i)}$ for all $1 \leq i \leq k$. Thus $eC = \sum_{i=1}^{k} \varepsilon_i A_{C_r(H_i)}$. But $xA_S = A_{xS}$ for any $x$ in $R$ and matrix $S$ and also $A_T + A_S = A_{T+S}$. Hence $eC = A_{(\sum_{i=1}^{k} \varepsilon_i C_r(H_i))}$. Thus every $g$-inverse of the form $eC$ is $A_S$ for some $S$. For this $S$ it is clear that $Tr(SC_r(A)) = e$.

♠

Over an integral domain every regular matrix $A$ is Rao-regular and its Rao-idempotent $I(A) = 1$ by Theorem 7.9 (b). Hence we have,

**Proposition 7.30:** Let $A$ be an $m \times n$ regular matrix over an integral domain with $\rho(A) = r$. Then every $g$-inverse of $A$ is of the form $A_S$ for some $\binom{n}{r} \times \binom{m}{r}$ matrix $S$ and for this $S, Tr(SC_r(A)) = 1$.

♠

Theorem 7.29 can be used to characterize all $g$-inverse of a regular matrix over a commutative ring.

**Theorem 7.31:** Let $A$ be an $m \times n$ regular matrix over a commutative ring $R$ with $\rho(A) = r > 0$. Let $A = e_1A + e_2A + \cdots + e_kA$ be the canonical decomposition of $A$ as in Theorem 7.19 with $\rho(e_iA) = r_i$ for $1 \leq i \leq k$. Then a matrix $G$ is a $g$-inverse of $A$ if and only if $G$ is of the form $A_{e_1S_1} + A_{e_2S_2} + \cdots + A_{e_kS_k} + (1 - e_1 - e_2 - \cdots - e_k)H$ where for $1 \leq i \leq k$ $e_iS_i$ is an $\binom{n}{r_i} \times \binom{m}{r_i}$ matrix with the property that $Tr(e_iC_{r_i}(S_iA)) = e_i$ and $H$ is some $n \times m$ matrix.

**Proof:** This is only a combination of Theorem 7.19 and Theorem 7.29.

♠

## 7.6 M-P inverses over commutative rings

A combination of the characterization of Rao-regular matrices which have Moore-Penrose inverses (Proposition 7.15) and the decomposition Theorem of section 7.4 gives us a characterization of matrices over commutative rings which have Moore-Penrose inverses.

**Theorem 7.32:** Let $R$ be a commutative ring with an involution $a \to \bar{a}$. Let $A$ be an $m \times n$ matrix over $R$ with Rao-index $t$, Rao-list of idempotents $(e_1, e_2, \cdots, e_t)$ and Rao-list of ranks $(r_1, r_2, \cdots, r_t)$. Then $A$ has Moore-Penrose inverse if and only if $e_tA = 0, \overline{e_i} = e_i$ for all $1 \leq i < t$ and $e_iA$ has Moore-Penrose inverse for each $i$. Specifically, if $G_i$ is the Moore-Penrose inverse of $e_iA$ and $\overline{e_i} = e_i$ for all $1 \leq i < t$ and $e_tA = 0$ then $G = G_1 + G_2 + \cdots + G_{t-1}$ is the Moore-Penrose inverse of $A$.

**Proof:** Let $G$ be the Moore-Penrose inverse of $A$. By Proposition 7.22, since $G$ is a $g$-inverse of $A$, we have $e_tA = 0$. Hence $A = e_1A + e_2A + \cdots + e_{t-1}A$ is the canonical decomposition of $A$ as in Theorem 7.19 (ii).

By Theorem 7.20 we have that $AG = e_1AG + e_2AG + \cdots + e_{t-1}AG$ is the canonical decomposition of $AG$. Since $G$ is the Moore-Penrose inverse of $A$ we have that $(AG)^* = AG$. Hence we get $AG = (AG)^* = \overline{e_1}(AG)^* + \overline{e_2}(AG)^* + \cdots + \overline{e_{t-1}}(AG)^* = \overline{e_1}AG + \overline{e_2}AG + \cdots + \overline{e_{t-1}}AG$. Also, since for $1 \leq i < t$, $e_i$ is the identity of $\mathcal{I}_{r_i}(e_iAG)$ we have that $\overline{e_i}$ is the identity of $\mathcal{I}_{r_i}(\overline{e_i}(AG)^*) = \mathcal{I}_{r_i}(\overline{e_i}AG)$. Hence $AG = \overline{e_1}AG + \overline{e_2}AG + \cdots + \overline{e_{t-1}}AG$ is a canonical decomposition of $AG$. By the uniqueness of the canonical decomposition(Theorem 7.19) we have $\overline{e_i} = e_i$ for $1 \leq i < t$. Now, since $G$ is the Moore-Penrose inverse of $A$ it easily follows that $e_iG$ is the Moore-Penrose inverse of $e_iA$ for all $1 \leq i < t$.

Conversely, if $G_i$ is the Moore-Penrose inverse of $e_iA$, and $\overline{e_i} = e_i$ for all $1 \leq i < t$ and if $e_tA = 0$ then $e_iG_i = G_i$ for all $1 \leq i < t$. Also, it is easily verified that $G_1 + G_2 + \cdots + G_{t-1}$ is the Moore-Penrose inverse of $A$.

♠

By Proposition 7.15, the previous Theorem takes the following form.

**Theorem 7.33:** Let $R$ be a commutative ring with an involution $a \to \bar{a}$. Let $A$ be an $m \times n$ matrix over $R$ with Rao-index $t$, Rao-list of idempotents $(e_1, e_2, \cdots, e_t)$ and Rao-list of ranks $(r_1, r_2, \cdots, r_t)$. Then $A$ has Moore-Penrose inverse if and only if $e_tA = 0, \overline{e_i} = e_i$ for all $1 \leq i < t$ and for all $1 \leq i < t$ there exists $w_i$ in $R$ such that $e_iTr(C_{r_i}(A^*A))w_i = e_i$. In this case $A^+ = A_{S_1} + A_{S_2} + \cdots + A_{S_{t-1}}$ where $S_i = e_iw_iC_{r_i}(A^*)$ for $1 \leq i < t$.

♠

## 7.7 Group inverses over commutative rings

We shall follow the method of section 7.6 for Group inverses too.

**Theorem 7.34:** Let $R$ be a commutative ring. Let $A$ be an $m \times m$ matrix over $R$ with Rao-index $t$, Rao-list of idempotents $(e_1, e_2, \cdots, e_t)$ and

Rao-list of ranks $(r_1, r_2, \cdots, r_t)$. Then $A$ has group inverse if and only if $e_tA = 0$ and for all $1 \leq i < t$, $e_iA$ has group inverse.

**Proof:** If $G$ is the group inverse of $A$ then it is easily verified that $e_iG$ is the group inverse of $e_iA$ for all $1 \leq i < t$. If $G_i$ is the group inverse of $e_iA$ for $1 \leq i < t$ then $e_iG_i = G_i$ for all $1 \leq i < t$. Also it can be verified that $G_1 + G_2 + \cdots + G_{t-1}$ is the group inverse of $A$.

♠

By Proposition 7.17, the above Theorem takes the following form.

**Theorem 7.35:** Let $R$ be a commutative ring. Let $A$ be an $m \times m$ matrix over $R$ with Rao-index $t$, Rao-list of idempotents $(e_1, e_2, \cdots, e_t)$ and Rao-list of ranks $(r_1, r_2, \cdots, r_t)$. Then $A$ has group inverse if and only if $e_tA = 0$ and for all $1 \leq i < t$ there exists $w_i$ in $R$ such that $e_i(Tr(C_{r_i}(A)))^2 w_i = e_i$.

In this case

$$A^{\#} = A_{S_1} + A_{S_2} + \cdots + A_{S_{t-1}} \quad \text{where} \quad S_i = w_i e_i C_{r_i}(A) \quad \text{for} \quad 1 \leq i < t.$$

♠

Also, the group inverse of any matrix $A$ over a commutative ring is a polynomial in $A$.

**Proposition 7.36:** For any $m \times m$ matrix $A$ over a commutative ring $R$ if $A$ has a group inverse then the group inverse $A^{\#}$ is a polynomial in $A$.

**Proof:** This follows from Theorem 7.35 and Proposition 7.18.

♠

## 7.8 Drazin inverses over commutative rings

In our discussion on Rao-regular matrices we had refrained from discussing Drazin inverses for the simple reason that a matrix admitting Drazin inverse need not be a regular matrix. Many general statements made in section 6.7, in which Drazin inverses of matrices over integral domains are discussed, are not true for matrices over commutative rings. We shall now present a careful discussion of Drazin inverses of matrices over commutative rings, sometimes referring to results of section 6.7.

We shall first give a finer definition of Drazin inverse. Let $R$ be a commutative ring. Let $A$ and $G$ be $m \times m$ matrices over $R$. Let $k \geq 1$ be an integer. We shall say that $G$ is a $k$-Drazin inverse of $A$ if $G$ and $A$ satisfy

$$A^{k+1}G = A^k \qquad (1^k)$$

$$GAG = G \tag{2}$$
$$\text{and } AG = GA \tag{5}$$

If $G$ is a $k$-Drazin inverse of $A$ then $G$ is an $\ell$-Drazin inverse of $A$ for all $\ell \geq k$. Also, if $A$ has a $k$-Drazin inverse then $k \geq p$, the index of $A$. It is possible that a matrix may have a $k$-Drazin inverse but not $p$-Drazin inverse, where $p$ is the index of $A$. If the ring is an integral domain, by Proposition 6.22 every $k$-Drazin inverse is a $p$-Drazin inverse which was called Drazin inverse in section 6.7.

Let us look at the matrix $A = \begin{bmatrix} 1 & 1 \\ 1 & 1 \end{bmatrix}$ over the ring $\mathbb{Z}[6]$. $\rho(A) = 1$ and $A$ does not have a 1-Drazin inverse because $A^2 = \begin{bmatrix} 2 & 2 \\ 2 & 2 \end{bmatrix}$ and there is no matrix $G$ such that $= \begin{bmatrix} 2 & 2 \\ 2 & 2 \end{bmatrix} G = \begin{bmatrix} 1 & 1 \\ 1 & 1 \end{bmatrix}$. However $A$ has a 2-Drazin inverse. Note that $A^3 = \begin{bmatrix} 4 & 4 \\ 4 & 4 \end{bmatrix}$ and $G = \begin{bmatrix} 4 & 4 \\ 4 & 4 \end{bmatrix}$ satisfies $A^3G = A^2, GAG = G$ and $AG = GA$. Thus $G$ is a 2-Drazin inverse of $A$.

Hence we shall study the existence of $k$-Drazin inverses. If a matrix $A$ has a $k$-Drazin inverse then the $k$-Drazin inverse is unique. The proof given for Proposition 6.23 holds good for the case of general commutative rings too.

The existence of $k$-Drazin inverse of a matrix $A$ is intimately related to the existence of group inverse (of some other matrix). We shall see this now.

**Proposition 7.37:** Let $A$ be an $m \times m$ matrix over a commutative ring $R$ and let $k \geq 1$ be an integer. Then $A$ has $k$-Drazin inverse if and only if $A^k$ has group inverse.

**Proof:** Suppose that $G$ is a $k$-Drazin inverse of $A$. Then $A^{k+1}G = A^k$, $GAG = G$ and $AG = GA$. We shall show that $G^k$ is the group inverse of $A^k$. $A^k = A^{k+1}G = A^kGA = A^kG^kA^k$, $G^kA^kG^k = G^kAG = G^{k-1}GAG = G^k$ and $G^kA^k = GA = AG = A^kG^k$. Thus $G^k$ is the group inverse of $A^k$.

Conversely, if $H$ is the group inverse of $A^k$ let us show that $A^{k-1}H$ is the $k$-Drazin inverse of $A$. Let $G = A^{k-1}H$. Then $A^{k+1}G = A^{k+1}A^{k-1}H = A^kHA^k = A^k$ and $GAG = A^{k-1}HAA^{k-1}H = A^{k-1}HA^kH = A^{k-1}H = G$. $H$ being the group inverse of $A^k$, by Proposition 7.36 $H$ is a polynomial in $A^k$. Hence $G = A^{k-1}H$ is a polynomial in $A$. Hence $AG = GA$. Thus $G$ is the $k$-Drazin inverse of $A$.

♠

It also follows from the above proof that the $k$-Drazin inverse of a matrix $A$, when it exists, is a polynomial in $A$. This is the analogue of Proposition 6.27 for matrices over commutative rings.

## EXERCISES:

**Exercise 7.1:** If $A$ is an $n \times n$ matrix over a commutative ring $R$ with $\rho(A) = 1$, show that $A^2 = Tr(A)A$. Show in general that $A^n = (Tr(A))^{n-1}A$ and $Tr(A^n) = (Tr(A))^n$ for all $n \geq 1$.

**Exercise 7.2:** For a matrix $B$ over a commutative ring with $\rho(B) = r$, if $\rho(B^2) = \rho(B)$ show that the sum of all $r \times r$ principal minors of $B$ $(= Tr(C_r(B)))$ is nonzero. Show that the converse is true if $R$ is an integral domain and that the converse need not be true if $R$ has zero divisors.

**Exercise 7.3:** Look for conditions for the existence of $\{1, 2, 3\}$-inverses of rank one matrices over commutative rings.

**Exercise 7.4:** Even over a regular commutative ring $R$ show that not every matrix need be Rao-regular, even though every matrix is regular ($\mathbb{Z}[6]$ is a good example). Characterize commutative rings $R$ over which every regular matrix is Rao-regular.

**Exercise 7.5:** Find necessary and sufficient conditions for a matrix over the ring of all real valued continuous functions on a normal topological space to be regular.

**Exercise 7.6:** Over the ring $R[x_1, x_2, \cdots, x_n]^*$ of Chapter 1 (10) show that a matrix $A$ is regular if and only if $A$ has a Moore-Penrose inverse if and only if $A$ has constant rank over all $(x_1, x_2, \cdots, x_n)$. See [94].

**Exercise 7.7:** Formulate and prove the exact analogue of Theorem 6.28 for matrices over commutative rings.

**Exercise 7.8:** For real matrices, if $A$ and $B$ have the same class of matrices as $g$-inverses then $A = B$. Examine this question for matrices over commutative rings. First examine this problem for matrices over integral domains. Clearly one needs to have that $A$ and $B$ are regular, because it is perfectly conceivable that $A \neq B$ and both $A$ and $B$ are not regular.

**Exercise 7.9:** Over a commutative ring, if $A$ is an $m \times m$ matrix, the sequence of ranks $(\rho(A^k))_{k \geq 1}$ does satisfy $\rho(A) \geq \rho(A^2) \geq \rho(A^3) \geq \cdots$ but does not behave as well as in the case of matrices over integral domains. Show that over commutative rings $\rho(A^k) = \rho(A^{k+1})$ need not imply that $\rho(A^{k+1}) = \rho(A^{k+2})$. In fact, show that any sequence of nonnegative integers $\ell_1 \geq \ell_2 \geq \cdots$ is a sequence of the type $\rho(A), \rho(A^2), \cdots$ for some matrix $A$ over a suitable commutative ring $R$.

**Exercise 7.10:** If $A = (a_{ij})$ is an $m \times n$ matrix over a commutative ring $R$ such that each $a_{ij}$ is regular in $R$ then can we conclude that $A$ is regular? i.e., Can we weaken the hypothesis of Proposition 7.25? Consider the matrix $\begin{bmatrix} 1 & 1 \\ 3 & 5 \end{bmatrix}$ over $\mathbb{Z}[12]$. However if either $m = 1$ or $n = 1$ show that we can conclude that $A$ is regular.

# Chapter 8

# Special topics

In this Chapter we shall look at various topics related to regularity of matrices over rings, several of these topics have been the subject of intense research.

## 8.1 Generalized Cramer Rule

If an $n \times n$ real matrix $A$ has an inverse and if $A\underline{x} = \underline{a}$ has to be solved then one sees that $\underline{x} = A^{-1}\underline{a}$ is a solution of $A\underline{x} = \underline{a}$. Also, the solution $\underline{x} = A^{-1}\underline{a}$ can be computed using Cramer rule, namely, if $\underline{x} = (x_1, x_2, \cdots, x_n)^T$ then $x_j = \frac{|A(j \to \underline{a})|}{|A|}$ for $1 \leq j \leq n$ where $A(j \to \underline{a})$ stands for the matrix obtained from $A$ by replacing the $j$th column of $A$ by $\underline{a}$. We shall generalize this method of finding solutions of equations to linear equations over commutative rings.

Let $A$ be an $m \times n$ matrix over a commutative ring $R$ with $\rho(A) = r$. Let $\underline{a}$ be an $m \times 1$ vector over $R$. If $1 \leq j \leq n$ we shall use the notation $A(j \to \underline{a})$ for the matrix obtained from $A$ by replacing the $j$th column of $A$ by $\underline{a}$.

**Proposition 8.1:** Let $A$ be an $m \times n$ matrix over a commutative ring $R$ with $\rho(A) = r$. Let $\underline{a} = (a_1, a_2, \cdots, a_m)^T$ be an $m \times 1$ vector over $R$. Let $S = (s_{\beta\alpha})_{\alpha \in Q_{r,m}, \beta \in Q_{r,n}}$ be an $\binom{n}{r} \times \binom{m}{r}$ matrix over $R$ and let $A_S = (g_{ij})$ be the $n \times m$ matrix defined by

$$g_{ij} = \sum_{\alpha \in Q_{r,m} \beta \in Q_{r,n}} \frac{\partial |A_{\alpha\beta}|}{\partial a_{ji}} s_{\beta\alpha}.$$

If $(x_1, x_2, \cdots, x_n)^T = \underline{x} = A_S\underline{a}$ then, for $1 \leq j \leq n$,

$$x_j = \sum_{\substack{\alpha:\alpha\in Q_{r,m} \\ \beta:j\in\beta,\beta\in Q_{r,n}}} |A(j \to \underline{a})_{\alpha\beta}| s_{\beta\alpha}.$$

**Proof:** For $1 \leq j \leq n, x_j = (A_S\underline{a})_j = \sum_k g_{jk}a_k$

$$= \sum_k \left( \sum_{\alpha\in Q_{r,m},\beta\in Q_{r,n}} \frac{\partial|A_{\alpha\beta}|}{\partial a_{kj}} s_{\beta\alpha} \right) a_k$$

$$= \sum_{\substack{\alpha:\alpha\in Q_{r,m} \\ \beta:j\in\beta,\beta\in Q_{r,n}}} \left( \sum_k \frac{\partial|A_{\alpha\beta}|}{\partial a_{kj}} a_k \right) s_{\beta\alpha}$$

$$= \sum_{\substack{\alpha:\alpha\in Q_{r,m} \\ \beta:j\in\beta,\beta\in Q_{r,n}}} |A(j \to \underline{a})_{\alpha\beta}| s_{\beta\alpha}.$$

Thus, for $1 \leq j \leq n$, $x_j$ is obtained as follows: Let $B$ be the matrix obtained from $A$ by replacing the $j$th column of $A$ by $\underline{a}$. Then for $1 \leq j \leq n$, set $x_j = \sum_{\substack{\alpha:\alpha\in Q_{r,m} \\ \beta:j\in\beta,\beta\in Q_{r,n}}} |B_{\alpha\beta}| s_{\beta\alpha}$.

♠

Now, we shall look at solutions of consistent system of equations for Rao-regular matrices.

**Proposition 8.2:** Let $A$ be an $m \times n$ Rao-regular matrix over a commutative ring $R$ with $\rho(A) = r$ and Rao-idempotent $I(A)$. Consider the consistent system of equations $A\underline{x} = \underline{a}$. If $I(A) = \sum_{(\alpha,\beta)\in\Delta} |A_{\alpha\beta}| c_{\beta\alpha}$ then a solution $\underline{x} = (x_1, x_2, \cdots, x_n)^T$ of $A\underline{x} = \underline{a}$ is given by

$$x_j = \sum_{\substack{(\alpha,\beta)\in\Delta \\ \beta:j\in\beta}} |A(j \to \underline{a})_{\alpha\beta}| c_{\beta\alpha}.$$

for $1 \leq j \leq n$.

**Proof:** This follows from Proposition 8.1 and Theorem 5.4.

♠

This method can be used to find minimum norm least square solutions for system of equations over real or complex numbers also. If one wants to find a minimum norm least square solution of $A\underline{x} = \underline{a}$, where $A$ is some

$m \times n$ nonzero real matrix and $\underline{a}$ is an $m \times 1$ vector then one has to find $A^+\underline{a}$ because $A^+\underline{a}$ is the minimum norm least square solution. But if $G$ is some $g$-inverse of $A$, by Proposition 7.15 $A^+ = A_S$ where $S = wC_r(A^*)$ and $Tr(C_r(A^*A))w = 1$. Now $A_S\underline{a}$ can be found using Proposition 8.1.

Also, note that, using Theorem 7.19 of representing any regular matrix in terms of Rao-regular matrices we can also find solutions of consistent $A\underline{x} = \underline{a}$ for a regular matrix $A$. If $A = e_1A + e_2A + \cdots + e_kA$ is the canonical decomposition of a regular matrix $A$ with Rao-regular matrices, $e_1A, e_2A, \cdots, e_kA$, then, for finding a solution of consistent $A\underline{x} = \underline{a}$, one simply uses Proposition 8.2 for $e_iA\underline{x} = e_i\underline{a}$ for $1 \le i \le k$ and finds the solution $\underline{x_i}$ of $e_iA\underline{x} = e_i\underline{a}$. Then $\underline{x_0} = \underline{x_1} + \cdots + \underline{x_k}$ is a solution of $A\underline{x} = \underline{a}$.

We shall call this method of finding a solution of a consistent system of equations $A\underline{x} = \underline{a}$ when $A$ is regular, the *Generalized Cramer Rule.* In case $A$ is an $n \times n$ unimodular matrix the Generalized Cramer Rule reduces to the well known Cramer Rule.

## 8.2 A rank condition for consistency

In the previous section we gave a method of finding solutions for consistent $A\underline{x} = \underline{a}$ when $A$ is regular. In this section we shall give a rank condition for consistency of $A\underline{x} = \underline{a}$ when $A$ is regular.

**Theorem 8.3 (Prasad):** For an $m \times n$ regular matrix $A$ over a commutative ring $R$ and an $m \times 1$ vector $\underline{a}$ over $R$, $A\underline{x} = \underline{a}$ is consistent if and only if $\rho(eA) = \rho(e[A : \underline{a}])$ for every idempotent $e$ of $R$ where $[A : \underline{a}]$ is the $m \times (n+1)$ augmented matrix.

**Proof:** If $A\underline{x} = \underline{a}$ is consistent then $\rho(eA) \le \rho(e[A : a]) = \rho(e[A : A\underline{x}]) = \rho(eA[I : \underline{x}]) \le \rho(eA)$. Thus $\rho(eA) = \rho(e[A : \underline{a}])$ for all idempotents $e$ of $R$.

To show the converse, let $\rho(A) = r$ and let $\rho(eA) = \rho(e[A : \underline{a}])$ for all idempotents $e$ of $R$. Hence $\rho([A : \underline{a}]) = r$.

Let us first assume that $A$ is Rao-regular with Rao-idempotent $I(A)$. Since $1 - I(A)$ is an idempotent, we have $0 = \rho((1 - I(A))A) = \rho((1 - I(A))[A : \underline{a}]$. Hence $I(A)[A : \underline{a}] = [A : \underline{a}]$. Also $I(A) \in \mathcal{I}_r(A) \subset \mathcal{I}_r([A : a])$.

Hence $[A : \underline{a}]$ is Rao-regular and its Rao-idempotent $I([A : \underline{a}]) = I(A)$. If $I(A) = \sum_{(\alpha,\beta)\in\Delta_{r,m,n}} |A_{\alpha\beta}|c_{\beta\alpha}$, then by taking

$$\begin{aligned} d_{\beta\alpha} &= c_{\beta\alpha} \quad \text{for } (\alpha,\beta) \in \Delta_{r,m,n} \text{ and} \\ &= 0 \quad \text{for } (\alpha,\beta) \in (\Delta_{r,m,n+1} \backslash \Delta_{r,m,n}) \end{aligned}$$

we see that $\sum_{(\alpha,\beta)\in\Delta_{r,m,n}} |[A : \underline{a}]_{\alpha\beta}| d_{\beta\alpha} = I(A).$

If we write $C = (c_{\beta\alpha})_{(\alpha,\beta)\in\Delta_{r,m,n}}$ and $D = (d_{\beta\alpha})_{(\alpha,\beta)\in\Delta_{r,m,n+1}}$ then $A_C$ as defined in section 5.3 is a generalized inverse of $A$, $[A : \underline{a}]_D$ as defined in section 5.3 is a generalized inverse of $[A : \underline{a}]$ and from the definition of $[A : \underline{a}]_D$ we have $[A : \underline{a}]_D = \begin{bmatrix} A_C \\ 0 \end{bmatrix}$. Hence $[A : \underline{a}] = [A : \underline{a}] \begin{bmatrix} A_C \\ 0 \end{bmatrix} [A : \underline{a}] = [AA_CA : AA_C\underline{a}] = [A : AA_C\underline{a}]$. Thus $AA_C\underline{a} = \underline{a}$, or $A\underline{x} = \underline{a}$ has a solution.

In case $A$ is regular, by the canonical decomposition Theorem of Prasad (Theorem 7.19 ) we can write $A = e_1A + \cdots + e_kA$ where $e_1A, e_2A, \cdots, e_kA$ are all Rao-regular. From the hypothesis, we get that $\rho(ee_iA) = \rho(ee_i[A : \underline{a}])$ for $i = 1, 2, \cdots, k$ and any idempotent $e$ of $R$. Hence from what was shown already for Rao-regular matrices we have that $e_iA\underline{x} = e_i\underline{a}$ has a solution for $i = 1, 2, \cdots, k$. If $\underline{x}_i$ for $i = 1, 2, \cdots, k$ are solutions for $e_iA\underline{x} = e_i\underline{a}$ for $i = 1, 2, \cdots, k$ respectively, then $\underline{x} = e_1\underline{x}_1 + e_2\underline{x}_2 + \cdots + e_k\underline{x}_k$ is a solution of $A\underline{x} = \underline{a}$.

♠

Thus it seems reasonable to define the *rank function* of a matrix $A$ over a commutative ring $R$ to be a nonnegative integer valued function $\rho_A$ defined on the idempotents of $R$ defined by $\rho_A(e) = \rho(eA)$ for idempotents $e$ of $R$. The above result says that if $A$ is regular $A\underline{x} = \underline{a}$ has solution if and only if $\rho_A = \rho_{[A:\underline{a}]}$. If $A$ is not regular even if $\rho_A = \rho_{[A:\underline{a}]}$, $A\underline{x} = \underline{a}$ may not be solvable. As an example take $A = \begin{bmatrix} 2 \\ 4 \end{bmatrix}$ and $\underline{a} = \begin{bmatrix} 3 \\ 6 \end{bmatrix}$ as matrices over the ring of integers.

Over the ring $\mathbb{Z}[k]$ for an integer $k$, solvability of $A\underline{x} = \underline{a}$ for a regular matrix can be treated using their rank functions to advantage.

## 8.3 Minors of reflexive $g$-inverses

In Theorem 6.2 we have seen that if $G = (g_{ij})$ is a reflexive $g$-inverse of an $m \times n$ matrix $A$ over an integral domain $R$ with $\rho(A) = r$ then $G = A_{C_r(G)}$, i.e., $g_{ij} = \sum_{(\alpha,\beta)\in\Delta} \frac{\partial|A_{\alpha\beta}|}{\partial a_{ji}} |G_{\beta\alpha}|$.

If $A$ is a matrix over a commutative ring $R$, the above result need no longer be true. Let us take the matrix $A = \begin{bmatrix} 1 & 0 \\ 0 & 4 \end{bmatrix}$ over the ring $\mathbb{Z}[12]$. Then $A$ is an idempotent matrix with $\rho(A) = 2$. Hence $A$ $(= G$, say $)$ is itself a reflexive $g$-inverse of $A$. But $C_r(G)$ is a $1 \times 1$ matrix, namely, 4 and $A_{C_r(G)} = \begin{bmatrix} 4 & 0 \\ 0 & 4 \end{bmatrix}$. Thus $A_{C_r(G)} \neq G$.

In the present section we shall generalize Theorem 6.2 to Rao-regular matrices over commutative rings. We shall also describe a method of obtaining the minors of reflexive $g$-inverses $G$ in terms of $C_r(G)$. We shall apply these results to the Moore-Penrose inverse and Group inverse and obtain generalizations of Jacobi's identity.

We shall start with a generalization of Theorem 5.4 (a).

**Proposition 8.4:** Let $A$ be an $m \times n$ matrix over a commutative ring $R$ with $\rho(A) = r$. Let $1 \leq p \leq r$. If $C = (c_{\beta\alpha})_{(\alpha,\beta)\in\Delta_{r,m,n}}$ is a matrix over $R$ such that $\left(\sum_{(\alpha,\beta)\in\Delta_{r,m,n}} |A_{\alpha\beta}|c_{\beta\alpha}\right) a_{k\ell} = a_{k\ell}$ for all $k$ and $\ell$, then the $\binom{n}{p} \times \binom{m}{p}$ matrix $D = (d_{\gamma\delta})_{\gamma\in Q_{p,n},\delta\in Q_{p,m}}$ defined by $d_{\gamma\delta} = \sum_{(\alpha,\beta)\in\Delta_{r,m,n}} \frac{\partial|A_{\alpha\beta}|}{\partial|A_{\delta\gamma}|}c_{\beta\alpha}$ is a $g$-inverse of the compound matrix $C_p(A)$. Moreover, if $\rho(C) = 1$ then $D$ is a reflexive $g$-inverse of $C_p(A)$.

**Proof:** For fixed $\alpha \in Q_{r,m}$, $\beta \in Q_{r,n}$, $\theta \in Q_{p,m}$ and $\eta \in Q_{p,n}$ consider the partitioned matrix

$$T = \begin{bmatrix} A_{\alpha\beta} & A_{\alpha\eta} \\ A_{\theta\beta} & A_{\theta\eta} \end{bmatrix}.$$

Since $\rho(A) = r$ we have that $\rho(T) \leq r$. Hence $|T| = 0$. By using the Laplace Expansion Theorem with respect to the last $p$ rows and the last $p$ columns of $T$ we see that

$$0 = |T| = |A_{\theta\eta}||A_{\alpha\beta}| - \sum_{\substack{\gamma\in Q_{p,n},\delta\in Q_{p,m} \\ \delta\subset\alpha,\gamma\subset\beta}} |A_{\theta\gamma}||A_{\delta\eta}|\frac{\partial|A_{\alpha\beta}|}{\partial|A_{\delta\gamma}|}.$$

While checking this, one should Exercise caution and also realize that $\delta \subset \alpha$ and $\gamma \subset \beta$.

Hence $|A_{\theta\eta}||A_{\alpha\beta}| = \sum_{\substack{\gamma\in Q_{p,n},\delta\in Q_{p,m} \\ \delta\subset\alpha,\gamma\subset\beta}} |A_{\theta\gamma}||A_{\delta\eta}|\frac{\partial|A_{\alpha\beta}|}{\partial|A_{\delta\gamma}|}$.

Now, let us see that $C_p(A)DC_p(A) = C_p(A)$. In fact,

$$\sum_{\gamma\in Q_{p,n},\delta\in Q_{p,m}} |A_{\theta\gamma}|d_{\gamma\delta}|A_{\delta\eta}|$$

$$= \sum_{\gamma\in Q_{p,n},\delta\in Q_{p,m}} |A_{\theta\gamma}||A_{\delta\eta}| \left(\sum_{(\alpha,\beta)\in\Delta_{r,m,n}} \frac{\partial|A_{\alpha\beta}|}{\partial|A_{\delta\gamma}|}c_{\beta\alpha}\right)$$

$$= \sum_{(\alpha,\beta)\in\Delta_{r,m,n}} c_{\beta\alpha} \left(\sum_{\substack{\gamma\in Q_{p,n},\delta\in Q_{p,m} \\ \delta\subset\alpha,\gamma\subset\beta}} |A_{\theta\gamma}||A_{\delta\eta}|\frac{\partial|A_{\alpha\beta}|}{\partial|A_{\delta\gamma}|}\right)$$

$$= \sum_{(\alpha,\beta)\in\Delta_{r,m,n}} c_{\beta\alpha}|A_{\theta\eta}||A_{\alpha\beta}|$$

$$= \left(\sum_{(\alpha,\beta)\in\Delta_{r,m,n}} |A_{\alpha\beta}|c_{\beta\alpha}\right)|A_{\theta\eta}| = |A_{\theta\eta}|.$$

The last equality follows from the hypothesis that

$$\left(\sum_{(\alpha,\beta)\in\Delta_{r,m,n}} |A_{\alpha\beta}|c_{\beta\alpha}\right) a_{k\ell} = a_{k\ell} \text{ for all } k \text{ and } \ell.$$

Thus $D$ is a $g$-inverse of $C_p(A)$.

The second part, namely, that $D$ is a reflexive $g$-inverse of $C_p(A)$ if $\rho(C) = 1$, though a little laborious, can be proved similarly, as in Theorem 5.5 (b) or Theorem 6.3.

♠

We need a result on group inverses which generalizes the last part of Proposition 7.17.

**Proposition 8.5:** Let $A$ be an $n \times n$ Rao-regular matrix over a commutative ring $R$ with $\rho(A) = r$ and Rao-idempotent $I(A)$. Let $A$ have group inverse and let $v$ be an element of $R$ such that $Tr(C_r(A))v = I(A)$. Let $1 \le p \le r$. Let $D = (d_{\gamma\delta})_{(\gamma,\delta)\in\Delta_{p,m,n}}$ be defined by

$$d_{\gamma\delta} = v^2 \sum_{(\alpha,\beta)\in\Delta_{r,m,n}} \frac{\partial|A_{\alpha\beta}|}{\partial|A_{\delta\gamma}|}|A_{\beta\alpha}|.$$

Then $D$ is the group inverse of $C_p(A)$.

**Proof:** This can be proved on the same lines as that of Proposition 7.17.

♠

We are now prepared for the main Theorem of this section. This generalizes Theorem 7.26.

**Theorem 8.6:** Let $A$ be an $m \times n$ Rao-regular matrix with $\rho(A) = r$ and Rao-idempotent $I(A)$. Let $G = (g_{ij})$ be a reflexive $g$-inverse of $A$.

Then, for $1 \le p \le r, \gamma \in Q_{p,n}$ and $\delta \in Q_{p,m}$,

$$|G_{\gamma\delta}| = \sum_{(\alpha,\beta)\in\Delta_{r,m,n}} \frac{\partial|A_{\alpha\beta}|}{\partial|A_{\delta\gamma}|}|G_{\beta\alpha}|.$$

**Proof:** This can be proved as in the case of the proof of Theorem 7.26 with the help of Proposition 8.5.

♠

We shall now use the above Theorem to find the minors of the Moore-Penrose inverse and the group inverse of a Rao-regular matrix when they exist. These results generalize the *Jacobi identity* which states that for an $n \times n$ nonsingular real matrix $A$, the $(\alpha, \beta)$-minor of $A^{-1}$, for $\alpha \in Q_{p,n}, \beta \in Q_{p,n}, 1 \leq p \leq n$ is given by

$$|(A^{-1})_{\alpha\beta}| = \frac{1}{|A|}\frac{\partial|A|}{\partial|A_{\alpha\beta}|}.$$

Thus the minors of $A^{-1}$ are related to the minors of $A$ by this identity.

We need two results which characterize $C_r(A^+)$ and $C_r(A^\#)$ for Rao-regular matrices. In fact, $C_r(A^+)$ and $C_r(A^\#)$ can be found even without finding $A^+$ and $A^\#$.

**Proposition 8.7:** Let $R$ be a commutative ring with an involution $a \to \bar{a}$. Let $A$ be an $m \times n$ Rao-regular matrix over $R$ with $\rho(A) = r$ and Rao-idempotent $I(A)$. Let $A^+$ exist. Let $w$ be an element of $R$ such that $\bar{w} = w$ and $Tr(C_r(A^*A))w = I(A)$. Then $C_r(A^+) = wC_r(A^*)$.

**Proof:** $C_r(A^+)$ is clearly the Moore-Penrose inverse of $C_r(A)$. Also, by Proposition 7.11 $C_r(A)$ is Rao-regular and $I(C_r(A)) = I(A)$. Since $C_r(A)$ is a rank one matrix, by Proposition 7.4, $(C_r(A))^+ = w(C_r(A))^* = wC_r(A^*)$ where $w$ is such that $\bar{w} = w$ and $Tr((C_r(A)^*)(C_r(A)))w = I(A)$. By the uniqueness of the Moore-Penrose inverse we have $C_r(A^+) = wC_r(A^*)$.

♠

The corresponding result for group inverse is the following.

**Proposition 8.8:** Let $A$ be an $n \times n$ Rao-regular matrix over a commutative ring with $\rho(A) = r$ and Rao-idempotent $I(A)$. Let $A^\#$ exist. Let $w$ be an element of $R$ such that $(Tr(C_r(A)))^2 w = I(A)$. Then $C_r(A^\#) = wC_r(A)$.

**Proof:** This can be proved exactly as in the previous Proposition using the fact that the group inverse of a matrix when it exists is unique.

♠

The above two Propositions supplement the results of Propositions 7.15 and 7.17 respectively. We shall now characterize the minors of the Moore-Penrose inverse of a Rao-regular matrix.

**Theorem 8.9:** Let $R$ be a commutative ring with an involution $a \to \bar{a}$. Let $A$ be an $m \times n$ Rao-regular matrix over $R$ with $\rho(A) = r$ and Rao-idempotent $I(A)$. Let $A^+$ exist. Let $w$ be an element of $R$ such that $\bar{w} = w$ and $Tr(C_r(A^*A))w = I(A)$. Let $1 \le p \le r$, $\gamma \in Q_{p,n}$ and $\delta \in Q_{p,m}$. Then

$$|(A^+)_{\gamma\delta}| = w \sum_{\substack{(\alpha,\beta)\in\Delta_{r,m,n} \\ \alpha:\delta\subset\alpha \\ \beta:\gamma\subset\beta}} \frac{\partial|A_{\alpha\beta}|}{\partial|A_{\delta\gamma}|}|A^*_{\beta\alpha}|.$$

In particular, if $R$ is an integral domain,

$$|(A^+)_{\gamma\delta}| = \frac{1}{Tr(C_r(A^*A))} \sum_{\substack{(\alpha,\beta)\in\Delta_{r,m,n} \\ \alpha:\delta\subset\alpha \\ \beta:\gamma\subset\beta}} \frac{\partial|A_{\alpha\beta}|}{\partial|A_{\delta\gamma}|}|A^*_{\beta\alpha}|.$$

**Proof:** $A^+$ being a reflexive $g$-inverse of $A$, by Theorem 8.6 $|(A^+)_{\gamma\delta} = \sum_{(\alpha,\beta)\in\Delta_{r,m,n}} \frac{\partial|A_{\alpha\beta}|}{\partial|A_{\delta\gamma}|}|(A^+)_{\beta\alpha}|$. But by Proposition 8.7, since $C_r(A^+) = wC_r(A^*)$ we have $|(A^+)_{\beta\alpha}| = w|A^*_{\beta\alpha}|$. Hence

$$|(A^+)_{\gamma\delta}| = w \sum_{\substack{(\alpha,\beta)\in\Delta_{r,m,n} \\ \alpha:\delta\subset\alpha \\ \beta:\gamma\subset\beta}} \frac{\partial|A_{\alpha\beta}|}{\partial|A_{\delta\gamma}|}|A^*_{\beta\alpha}|.$$

Thus the result.

♠

The minors of the group inverse are given by,

**Theorem 8.10:** Let $A$ be an $n \times n$ Rao-regular matrix over a commutative ring $R$ with $\rho(A) = r$ and Rao-idempotent $I(A)$. Let $A^\#$ exist. Let $w$ be an element of $R$ such that $(Tr(C_r(A)))^2 w = I(A)$. Let $1 \le p \le r, \gamma \in Q_{p,n}$ and $\delta \in Q_{p,n}$. Then

$$|(A^\#)_{\gamma\delta}| = w \sum_{\substack{(\alpha,\beta)\in\Delta_{r,m,n} \\ \alpha:\delta\subset\alpha \\ \beta:\gamma\subset\beta}} \frac{\partial|A_{\alpha\beta}|}{\partial|A_{\delta\gamma}|}|A_{\beta\alpha}|.$$

**Proof:** This can be proved exactly as in the case of Theorem 8.9.

♠

## 8.4 Bordering of regular matrices

Every $m \times n$ matrix $A$ over a commutative ring with $\rho(A) = r$ can be completed, in a way, to a unimodular matrix. In fact, $\begin{bmatrix} A & I_m \\ I_n & 0 \end{bmatrix}$ where $I_k$ is the $k \times k$ identity matrix, is a unimodular matrix. As is normal, any result which one gets without any effort is uninteresting. We shall look for a better result.

If $P$, $Q$ and $S$ are matrices over $R$ such that $\begin{bmatrix} A & P \\ Q & S \end{bmatrix}$ is unimodular then $P$ must be of order $m \times k$ with $k \geq (m-r)$ and $Q$ must be of order $\ell \times n$ with $\ell \geq (n-r)$. Hence we look for unimodular $\begin{bmatrix} A & P \\ Q & S \end{bmatrix}$ with minimum possible $k = m - r$ and minimum possible $\ell = n - r$. It turns out that if there is such a unimodular $\begin{bmatrix} A & P \\ Q & S \end{bmatrix}$ then $A$ must be regular. Other useful results also follow.

If $A$ is an $m \times n$ matrix over a commutative ring $R$ with $\rho(A) = r$ then we shall call a matrix $T = \begin{bmatrix} A & P \\ Q & S \end{bmatrix}$ a *bordering* of $A$ if $P$ is of order $m \times (m-r)$, $Q$ is of order $(n-r) \times n$ and $T$ is unimodular.

Let us recall that a commutative ring $R$ is called a projective free ring if every finitely generated projective module over $R$ is free. We have seen in Theorem 4.4 that over a principal ideal ring, if $g.c.d.(a_1, a_2, \cdots, a_n) = 1$ then there is an $n \times n$ unimodular matrix $A$ such that $(a_1, a_2, \cdots, a_n)$ is the first row of $A$. On the same lines, in section 4.6 we have seen that over a projective free integral domain every $m \times n$ matrix (with $m \leq n$) which has a right inverse can be completed to an $n \times n$ unimodular matrix. We shall generalize this result in this section: If $A$ is an $m \times n$ regular matrix over a projective free ring $R$ with $\rho(A) = r$ then there is an $m \times (m-r)$ matrix $P$ such that $[A \;\; P]$ has right inverse. We shall first investigate some necessary conditions for the existence of such a $P$. Recall that $BC$ is called a rank factorization of an $m \times n$ matrix $A$ with $\rho(A) = r$ if $BC = A$, $B$ is an $m \times r$ matrix with a left inverse and $C$ is an $r \times m$ matrix with a right inverse.

**Proposition 8.11:** Let $A$ be an $m \times n$ matrix over a commutative ring with $\rho(A) = r < m$. If there is an $m \times (m-r)$ matrix $P$ such that $T = [A \;\; P]$ has right inverse then $A$ is regular. If $\begin{bmatrix} G \\ V \end{bmatrix}$ is a right inverse of $T$ then $VA = 0$, $VP = I$, $G$ is a $g$-inverse of $A$ and $PV$ is a rank factorization of $I - AG$.

**Proof:** Let $T = (t_{ij})$. Since $T$ has right inverse, by Theorem 3.7, there exists $\{c_\beta\}_{\beta \in Q_{m,n+m-r}}$, elements of $R$ such that

$$\sum_{\beta \in Q_{m,n+m-r}} |T_{*\beta}| c_\beta = 1.$$

Since $\rho(A) = r, \rho([A \quad P]) = m$ and $P$ is of order $m \times (m-r)$ we get that $\rho(P) = m - r$. Also $|T_{*\beta}|$ can be nonzero only if $\{n+1, n+2, \cdots, n+m-r\} \subset \beta$. Let $\beta' = \beta - \{n+1, n+2, \cdots, n+m-r\}$ if $|T_{*\beta}| \neq 0$. For $\gamma \in Q_{m-r,m}$, let $\bar{\gamma} = \{1, 2, \cdots, m\} \backslash \gamma$ and $sgn(\gamma) = (-1)^{\sum_{i=1}^{m-r} (\gamma_i + n + i)}$ where $\gamma = \{\gamma_1, \gamma_2, \cdots, \gamma_{n-r}\}$.

Now, if $|T_{*\beta}| \neq 0$ then

$$|T_{*\beta}| = \sum_{\gamma \in Q_{m-r,m}} sgn(\gamma) |P_{\gamma *}| |A_{\bar{\gamma}\beta'}|.$$

Hence,

$$\begin{aligned} 1 &= \sum_{\beta \in Q_{m,n+m-r}} |T_{*\beta}| c_\beta \\ &= \sum_{\substack{\beta \in Q_{m,n+m-r} \\ \beta : |T_{*\beta}| \neq 0}} \sum_{\gamma \in Q_{m-r,m}} sgn(\gamma) |P_{\gamma *}| |A_{\bar{\gamma}\beta'}| c_\beta \\ &= \sum_{\gamma \in Q_{m-r,m}} \left( \sum_{\substack{\beta \in Q_{m,n+m-r} \\ \beta : |T_{*\beta}| \neq 0}} sgn(\gamma) |A_{\bar{\gamma}\beta'}| c_\beta \right) |P_{\gamma *}|. \end{aligned}$$

Hence a linear combination of all the $(m-r) \times (m-r)$ minors of $P$ is equal to one. By Theorem 3.7, $B = (b_{ij})$ defined by

$$\begin{aligned} b_{ij} &= \sum_{\gamma \in Q_{m-r,m}} \left( \sum_{\substack{\beta \in Q_{m,n+m-r} \\ \beta : |T_{*\beta}| \neq 0}} sgn(\gamma) |A_{\bar{\gamma}\beta'}| c_\beta \right) \frac{\partial}{\partial p_{ji}} |P_{\gamma *}| \\ &= \sum_{\beta : |T_{*\beta}| \neq 0} c_\beta \frac{\partial |T_{*\beta}|}{\partial t_{j,n+i}} \end{aligned}$$

is a left inverse of $P$, i.e., $BP = I$.

Now,

$$\begin{aligned}(BA)_{ik} &= \sum_j \left( \sum_{\beta:|T_{*\beta}|\neq 0} c_\beta \frac{\partial |T_{*\beta}|}{\partial t_{j,n+i}} \right) a_{jk} \\ &= \sum_{\beta:|T_{*\beta}|\neq 0} |S_{*\beta}| c_\beta\end{aligned}$$

where $S$ is the matrix obtained from $T$ by replacing the $(n+i)$th column of $T$ by the $k$th column of $A$. But then $\rho(S) < m$. Hence $|S_{*\beta}| = 0$ for every $\beta \in Q_{m,m+n-r}$. Hence $BA = 0$. Now, let $\begin{bmatrix} G \\ V \end{bmatrix}$ be a right inverse of $[A \;\; P]$. Then $AG + PV = I$. Pre-multiplying by $B$, we get that $BAG + BPV = B$. Hence $V = B$. Thus $VA = 0$ and $VP = I$. Again, by post-multiplying $AG + PV = I$ by $A$, we get $AGA + PVA = A$. Hence $AGA = A$. Thus $A$ is regular.

♠

A similar result also holds for matrices $\begin{bmatrix} A \\ Q \end{bmatrix}$ where $A$ is an $m \times n$ matrix with $\rho(A) = r$, $Q$ is an $(n-r) \times n$ matrix and $\begin{bmatrix} A \\ Q \end{bmatrix}$ has a left inverse. This and the above Proposition give us a remarkable consequences on Borderings of matrices.

**Proposition 8.12:** Let $A$ be an $m \times n$ matrix over a commutative ring with $\rho(A) = r$. If $T = \begin{bmatrix} A & P \\ Q & R \end{bmatrix}$ is a bordering of $A$ then $A$ is regular. If $T^{-1} = \begin{bmatrix} G & U \\ V & W \end{bmatrix}$ then $W = 0$, $PV$ is a rank factorization of $I - AG$, $UQ$ is a rank factorization of $I - GA$, $G$ is a $g$-inverse of $A$ and $R = -QGP$. Conversely, if $A$ is regular, $G$ is a reflexive $g$-inverse of $A$, $I - AG$ has a rank factorization $PV$ and $I - GA$ has a rank factorization $UQ$ then $T = \begin{bmatrix} A & P \\ Q & O \end{bmatrix}$ is a bordering of $A$ with $T^{-1} = \begin{bmatrix} G & U \\ V & O \end{bmatrix}$.

**Proof:** Let $T = \begin{bmatrix} A & P \\ Q & R \end{bmatrix}$ be a bordering of $A$ with $T^{-1} = \begin{bmatrix} G & U \\ V & W \end{bmatrix}$. Then $\begin{bmatrix} G \\ V \end{bmatrix}$ is a right inverse of $[A \;\; P]$ and $\begin{bmatrix} A \\ Q \end{bmatrix}$ is a right inverse of $[G \;\; U]$. Also, $P$ is an $m\times(m-r)$ matrix and $U$ is an $n\times(n-r)$ matrix. By Proposition 8.11 we get that $VA = 0, VP = I, I - AG = PV, I - GA = UQ$ and that $G$ is a $g$-inverse of $A$. Hence $I - AG$ and $I - GA$ have rank factorizations. Also, $TT^{-1} = I$ gives us that $AU + PW = 0$. Pre-multiplying by $V$ gives us that $W = 0$. One easily sees that $R = -QGP$. The converse, namely that, if $AGA = A, GAG = G$ and $I - AG = PV$ and $I - GA = UQ$ are rank factorizations then $\begin{bmatrix} A & P \\ Q & O \end{bmatrix} \begin{bmatrix} G & U \\ V & O \end{bmatrix} = \begin{bmatrix} I & O \\ O & I \end{bmatrix}$ is easily verified.

♠

The previous Proposition can be used to give simple characterizations for the existence of bordering of a regular matrix. We shall first give necessary and sufficient conditions for a regular matrix to admit a rank factorization. Let us identify any $m \times n$ matrix $A$ over a commutative ring $R$ as a module homomorphism from $R^n$ to $R^m$ defined by $\underline{x} \to A\underline{x}$.

**Proposition 8.13:** Let $R$ be a commutative ring.

(a) An idempotent matrix $B$ over $R$ has rank factorization if and only if $Range(B)$ is a free module.

(b) A regular matrix $A$ over $R$ has rank factorization if and only if $Range(A)$ is a free module.

**Proof:** (a) Let $B$ be an $m \times m$ idempotent matrix and let $S = Range(B)$. Let $B = MN$ be a rank factorization of $B$. Let $M_L^{-1}$ be a left inverse of $M$ and let $N_R^{-1}$ be a right inverse of $N$. $B\underline{x} = MN\underline{x}$ and $M\underline{y} = MNN_R^{-1}\underline{y} = AN_R^{-1}\underline{y}$ show that $Range(B) = Range(M)$. The module homomorphism $M_L^{-1}$ from $R^m$ to $R^k$ is, when restricted to $Range(M)$ is both injective and surjective (to $R^m$). Hence $Range(B)$ is free.

Conversely, If $Range(B)$ is free let $\phi : Range(B) \to R^k$ be an isomorphism for some $k$. Let $i$ be the identity (inclusion) map from $Range(B)$ to $R^m$. Then $B = i\phi^{-1}\phi B$, and $\phi B : R^m \to R^k$ and $i\phi^{-1} : R^k \to R^m$ are module homomorphisms. If we call the matrices corresponding to $i\phi^{-1}$ and $\phi B$ as $M$ and $N$ then $B = MN$. Also observe that $NM = \phi B i\phi^{-1} = I$ because $B$ is identity on the $Range(B)$. Hence $M$ has a left inverse and $N$ has a right inverse. Thus $B$ has a rank factorization.

(b) If $A$ has rank factorization then the first part of the proof of (a) which is valid for any $m \times n$ matrix $A$ tells us that $Range(A)$ is free. If $A$ is an $m \times n$ regular matrix and if $G$ is a $g$-inverse of $A$ then $B = AG$ is an idempotent matrix and $Range(B) = Range(A)$. If $Range(A)$ is free then we have that $Range(B)$ is free. In the notation of the proof of (a) above we have $AG = i\phi^{-1}\phi AG$. Hence post multiplication by $A$ gives us $A = i\phi^{-1}\phi A$. If we let $S = i\phi^{-1}$ and $T = \phi A$ then $A = ST$ and $TGS = \phi AGi\phi^{-1} = I$. Thus $A = ST$ is a rank factorization of $A$.

♠

Now we shall give the characterization of bordering.

**Proposition 8.14:** For an $m \times n$ matrix $A$ the following are equivalent.

(i) $A$ has bordering.

(ii) $A$ is regular and there is a $g$-inverse $G$ of $A$ such that $I - AG$ and $I - GA$ have rank factorizations.

(iii) $A$ is regular and there is a $g$-inverse $G$ of $A$ such that $Range(I - AG)$ and $Range(I - GA)$ are free.

(iv) $A$ is regular and $Ker(A)$ and $Coker(A)$ are free.

(v) $A$ is regular and for every $g$-inverse $G$ of $A$, $I - AG$ and $I - GA$ have rank factorizations.

(vi) $A$ is regular and for every $g$-inverse $G$ of $A$, $Range(I - AG)$ and $Range(I - GA)$ are free.

**Proof:** (i) $\Rightarrow$ (ii) follows from Proposition 8.12. If for some $g$-inverse $G$ of $A$ (ii) is satisfied then for the reflexive $g$-inverse $H = GAG$, $I - AH = I - AG$ and $I - HA = I - GA$. Hence from Proposition 8.12 again (ii) $\Rightarrow$ (i) follows.

(ii) $\Leftrightarrow$ (iii) follows from Proposition 8.13 because $I - AG$ and $I - GA$ are both idempotent.

Let us see that if $G$ is a $g$-inverse of $A$ then $Ker(A) = Range(I - GA)$ and $Coker(A)$ is isomorphic to $Range(I - AG)$. If $\underline{x} \in Ker(A)$ then $(I - GA)\underline{x} = \underline{x}$ and if $(I - GA)\underline{y} = \underline{x}$ then $A\underline{x} = \underline{0}$. Thus $Ker(A) = Range(I - GA)$. Recall that $Coker(A)$ is the quotient module $R^m / Range(A)$. Also, $R^m$ is isomorphic to $Coker(A) \oplus Range(A)$. On the other hand the matrix $AG$ is an $m \times m$ idempotent matrix and hence $R^m = Range(AG) \oplus Range(I - AG)$. But since $G$ is a $g$-inverse of $A$, $Range(AG) = Range(A)$. Hence $Coker(A)$ is isomorphic to $Range(I - AG)$. Thus $Range(I - GA)$ and $Range(I - AG)$ are free if and only if $Ker(A)$ and $Coker(A)$ are free. Thus we have (iii) $\Leftrightarrow$ (iv).

Note that in the above paragraph $G$ can be any $g$-inverse of $A$. Hence we have (iv) $\Leftrightarrow$ (v) $\Leftrightarrow$ (vi).

♠

Now we shall characterize all those commutative rings over which every regular matrix has a bordering.

**Theorem 8.15:** The following are equivalent over any commutative ring $R$.

(i) $R$ is projective free.

(ii) Every regular matrix over $R$ has rank factorization.

(iii) Every idempotent matrix over $R$ has rank factorization.

(iv) Every regular matrix over $R$ has a bordering.

(v) For every regular $m \times n$ matrix $A$ over $R$ with $\rho(A) = r$ there is an $m \times (m - r)$ matrix $P$ such that $[A \quad P]$ has a right inverse.

(vi) Every regular matrix over $R$ is of the form $U \begin{bmatrix} I & 0 \\ 0 & 0 \end{bmatrix} V$ for some unimodular matrices $U$ and $V$.

(vii) Every idempotent matrix over $R$ is of the form $U \begin{bmatrix} I & 0 \\ 0 & 0 \end{bmatrix} U^{-1}$ for some unimodular matrix $U$.

**Proof:** (i) $\Rightarrow$ (ii). For any $m \times n$ matrix $A$, $R^m = Range(A) \oplus Coker(A)$. Hence $Range(A)$ is a finitely generated projective module. From (i) we have that $Range(A)$ is free. By Proposition 8.13, since $A$ is regular, we get that $A$ has rank factorization.

(ii) $\Rightarrow$ (iii) is clear because every idempotent matrix is regular.

For (iii) $\Rightarrow$ (iv), if $A$ is a regular matrix with a reflexive $g$-inverse $G$ then $I - GA$ and $I - AG$ are idempotent matrices and hence by (iii), have rank factorizations. By Proposition 8.12, $A$ has bordering.

For (iv) $\Rightarrow$ (v), observe that if $T = \begin{bmatrix} A & P \\ Q & O \end{bmatrix}$ is a bordering of $A$ then $[A \quad P]$ has a right inverse.

(v) $\Rightarrow$ (iv) can be easily obtained by applying (v) twice.

For (iv) $\Rightarrow$ (i), let $X$ be a finitely generated projective module with $X \oplus Y = R^n$. Let $A : R^n \to Y$ be the natural projection. Then $I - A$ is idempotent and $Range(I - A) = X$. Since $A$ is idempotent, $A$ is a $g$-inverse of itself and by Proposition 8.14 (i) $\Leftrightarrow$ (v) we have that $Range(I - A)$ has a rank factorization. Now, by Proposition 8.13 we have that $X$ is free.

Thus (i)$-$(v) are shown to be equivalent. Now let us show that these are equivalent to (vi). Let us assume (i)$-$(v). Let $A$ be a regular matrix. Let $A = LM$ be a rank factorization of $A$ given by (ii). Since $L$ is a matrix with a left inverse and $M$ is a matrix with a right inverse, by (v) there exist matrices $V$ and $W$ such that $[L \quad V]$ and $\begin{bmatrix} M \\ W \end{bmatrix}$ are both unimodular. Now it is easily verified that $A = [L \quad V] \begin{bmatrix} I & 0 \\ 0 & 0 \end{bmatrix} \begin{bmatrix} M \\ W \end{bmatrix}$.

(vi) $\Rightarrow$ (ii) is clear.

Let us now see that (vii) also follows from (i)$-$(v). Let $A$ be an $n \times n$ idempotent matrix. Then $I - A$ is also idempotent. Let $A = LM$ and $I - A = ST$ be rank factorizations of $A$ and $I - A$ respectively. Then clearly $ML = I$ and $TS = I$. If $r = \rho(A)$ then $r = Tr(ML) = Tr(LM) = Tr(A)$. Similarly, $\rho(I-A) = Tr(I-A)$. But $n = Tr(A) + Tr(I-A)$. Thus $\rho(I-A) = n - r$. Hence the matrices $[L \quad S]$ and $\begin{bmatrix} M \\ T \end{bmatrix}$ both are $n \times n$ matrices. Also,

$[L \quad S]\begin{bmatrix} M \\ T \end{bmatrix} = I$, $[L \quad S]\begin{bmatrix} I & 0 \\ 0 & 0 \end{bmatrix}\begin{bmatrix} M \\ T \end{bmatrix} = A$ and $U = [L \quad S]$ is a unimodular matrix.

(vii) $\Rightarrow$ (iii) is clear.

♠

A nice consequence of the above Theorem is the following

**Proposition 8.16:** Over a projective free ring every $m \times n$ regular matrix $A$ with $\rho(A) = m$ can be completed to an $n \times n$ unimodular matrix. Also, every $m \times n$ regular matrix $A$ with $\rho(A) = m$ must have a right inverse.

♠

## 8.5 Regularity over Banach algebras

A unital complex commutative Banach algebra is a nice example of a commutative ring and as such the results of Chapters 5 and 7 are applicable. But because of the rich theory of Banach algebras some of the results on regularity of matrices can be expressed in Banach algebraic language. We shall do this in this section.

We shall not give any details of the notions and results from the Theory of Banach algebras but refer the reader to [35] or any other book on Banach algebras.

The maximal ideal space of a unital complex commutative Banach algebra B will be denoted by $\mathcal{M}$. There is a one-one correspondence between $\mathcal{M}$ and all linear multiplicative homomorphisms $\phi : \mathcal{B} \to \mathbf{C}$, the field of complex numbers. If $M \in \mathcal{M}$, there is a linear multiplicative homomorphism $\phi : B \to \mathbf{C}$ such that $M = \phi^{-1}\{0\}$. We shall identify $\mathcal{M}$ with all the linear multiplicative homomorphism $\phi : B \to \mathbf{C}$. On $\mathcal{M}$ there is a natural topology. If $e \in \mathcal{B}$ is idempotent and $\phi \in \mathcal{M}$ then $\phi(e) = 0$ or 1. If $e \in \mathcal{B}$ is idempotent $F_e = \{\phi : \phi(e) = 1\}$ is a clopen (closed and open) subset of $\mathcal{M}$ and is nonempty if $e \neq 0$. If $e_1$ and $e_2$ are two idempotents such that $e_1 e_2 = 0$ then $F_{e_1} \cap F_{e_2} = \emptyset$.

An element $x$ of $\mathcal{B}$ is unimodular if and only if $\phi(x) \neq 0$ for every $\phi \in \mathcal{M}$. In the same way, for $x_1, x_2, \cdots, x_\ell$ in $\mathcal{B}$ there exists $y_1, y_2, \cdots, y_\ell$ in $\mathcal{B}$ such that $\sum_{i=1}^{\ell} x_i y_i = 1$ if and only if for every $\phi \in \mathcal{M}$ there is at least one $i$ such that $\phi(x_i) \neq 0$.

If $A = (a_{ij})$ is an $m \times n$ matrix over $\mathcal{B}$ and if $\phi \in \mathcal{M}$ we write $\hat{A}(\phi)$ for the matrix whose $(i,j)$th element is $\hat{a}_{ij}(\phi) = \phi(a_{ij})$. $\hat{A}(\phi)$ is a complex matrix and $\rho(\hat{A}(\phi)) \leq \rho(A)$ for all $\phi$.

We shall first characterize certain Rao-regular matrices over $\mathcal{B}$.

**Proposition 8.17:** Let $A$ be an $m \times n$ matrix over a unital complex commutative Banach algebra with $\rho(A) = r$. The following are equivalent.

(i) $A$ is Rao-regular with Rao-idempotent $I(A) = 1$.

(ii) $\rho(A) = \rho(\hat{A}(\phi))$ for all $\phi \in \mathcal{M}$.

(iii) for every $\phi \in \mathcal{M}$ there exists $(\alpha, \beta) \in \Delta_{r,m,n}$ such that $\phi(|A_{\alpha\beta}|) \neq 0$.

(iv) There exists $\{c_{\beta\alpha} : (\alpha, \beta) \in \Delta\}$ such that $\sum_{(\alpha,\beta)\in\Delta_{r,m,n}} |A_{\alpha\beta}|c_{\beta\alpha} = 1$.

**Proof:** (iv) is a restatement of (i). (ii) $\Leftrightarrow$ (iv) is an elementary result in the theory of Banach algebras as was stated above. If $r = \rho(A)$ and if $\phi \in \mathcal{M}$, by (iii) there exists $(\alpha, \beta) \in \Delta$ such that $\phi(|A_{\alpha\beta}|) \neq 0$. Hence $|\hat{A}_{\alpha\beta}(\phi)| \neq 0$. Thus $\rho(\hat{A}(\phi)) \geq r$. Hence $\rho(A) = \rho(\hat{A}(\phi))$ for all $\phi \in \mathcal{M}$. Thus (iii) $\Rightarrow$ (ii) is proved. But (ii) $\Rightarrow$ (iii) is clear. Thus the Proposition is proved.

♠

Now that we have a characterization of Rao-regular matrices we can characterize regular matrices over Banach algebras.

**Theorem 8.18:** An $m \times n$ matrix $A$ over a unital complex commutative Banach algebra $\mathcal{B}$ is regular if and only if there exists a pairwise orthogonal set of idempotents $\{e_1, e_2, \cdots, e_k\}$ such that

(a) $A = e_1A + e_2A + \cdots + e_kA$;

(b) $\rho(e_1(A) > \rho(e_2A) > \cdots > \rho(e_kA) > 0$; and

(c) for every $1 \leq i \leq k, \rho(e_iA) = \rho(e_i\hat{A}(\phi))$ for all $\phi \in F_{e_i}$.

**Proof:** From the canonical decomposition of Prasad (Theorem 7.19) we get the orthogonal set of idempotents satisfying (a) and (b) of Theorem 7.19 (ii). For $1 \leq i \leq k, e_iA$ is Rao-regular. If we consider the Banach algebra $e_i\mathcal{B} = \{e_ix : x \in \mathcal{B}\}$ then $e_i$ is the 1 of $e_i\mathcal{B}$ and $e_iA$ is a Rao-regular matrix over $e_i\mathcal{B}$ with Rao-idempotent $I(e_iA) = e_i$, the 1 of $e_i\mathcal{B}$. Hence the previous Proposition completes the proof. The converse is also clear using the previous Proposition.

♠

All the results on Moore-Penrose inverses, group inverses and Drazin inverses from Chapter 7 will hold for matrices over Banach algebras too. We shall express some of these results in terms of $\phi \in \mathcal{M}$. Consider a unital complex commutative Banach algebra with an involution $a \to \bar{a}$. We shall call the involution a *symmetric involution* if $\phi(\bar{a}) = \overline{\phi(a)}$ for every $\phi \in \mathcal{M}$ where $\overline{\phi(a)}$ is the complex conjugate of $\phi(a)$.

**Theorem 8.19:** If $A$ is an $m \times n$ Rao-regular matrix with Rao-idempotent $I(A) = 1$ over a unital complex commutative Banach algebra $\mathcal{B}$ with a symmetric involution then $A$ has Moore-Penrose inverse. Also, over a unital complex commutative Banach algebra with a symmetric involution every regular matrix has Moore-Penrose inverse.

**Proof:** Let $\rho(A) = r$. According to Proposition 8.17 if $A$ is Rao-regular with Rao-idempotent $I(A) = 1$ then for every $\phi \in \mathcal{M}$ there exists $(\alpha, \beta) \in \Delta_{r,m,n}$ such that $\phi(|A_{\alpha\beta}|) \neq 0$. Hence for every $\phi \in \mathcal{M}$

$$\sum_{(\alpha,\beta)\in\Delta_{r,m,n}} \overline{\phi(|A_{\alpha\beta}|)}\phi(|A_{\alpha\beta}|) = \sum_{(\alpha,\beta)\in\Delta_{r,m,n}} \phi(\overline{|A_{\alpha\beta}|})\phi(|A_{\alpha\beta}|)$$
$$= \sum_{(\alpha,\beta)\in\Delta_{r,m,n}} \widehat{|A^*_{\beta\alpha}||A_{\alpha\beta}|}(\phi) \neq 0.$$

Hence $\sum_{(\alpha,\beta)\in\Delta_{r,m,n}} |A^*_{\beta\alpha}||A_{\alpha\beta}|$ is an invertible element of $\mathcal{B}$ because of a result mentioned at the beginning of this section. Thus, since $Tr(C_r(A^*A))$ is invertible, by Proposition 7.15 we have that $A$ has Moore-Penrose inverse.

The second part of the Theorem follows from Theorem 7.32.

♠

Regarding group inverses of matrices over Banach algebras we have

**Theorem 8.20:** If $A$ is an $n \times n$ Rao-regular matrix with Rao-idempotent $I(A) = 1$ over a unital complex commutative Banach algebra, $A$ has group inverse if and only if $\hat{A}(\phi)$ has group inverse for every $\phi \in \mathcal{M}$.

**Proof:** If $A$ has group inverse clearly $\hat{A}(\phi)$ has group inverse for every $\phi \in \mathcal{M}$.

Let $\rho(A) = r$. Since $A$ is Rao-regular by Proposition 8.17 we have $\rho(\hat{A}(\phi)) = r$ for all $\phi \in \mathcal{M}$. If $\hat{A}(\phi)$ has group inverse then $Tr(C_r(\hat{A}(\phi))) \neq 0$. Hence by the result mentioned at the beginning of this section, since $\sum_{\alpha\in Q_{r,n}} |A_{\alpha\alpha}|$ is unimodular in $\mathcal{B}$. Hence $Tr(C_r(A))|I(A)(= 1)$. By Proposition 7.17 we get that $A$ has group-inverse.

♠

For a general matrix we have

**Theorem 8.21:** Let $A$ be an $n \times n$ matrix over a unital complex commutative Banach algebra $\mathcal{B}$ with $\rho(A) = r$. Then the following are equivalent.

(i) $A$ has group inverse.

(ii) $A$ is regular and $\hat{A}(\phi)$ has group inverse for every $\phi \in \mathcal{M}$.

(iii) $A$ is regular and $\sum_{\alpha \in Q_{r,n}} |\widehat{A_{\alpha\alpha}}|(\phi) \neq 0$ for every $\phi \in \mathcal{M}$.

(iv) $A$ is regular and $\rho(\hat{A}(\phi)) = \rho(\widehat{A^2}(\phi))$ for every $\phi \in \mathcal{M}$.

**Proof:** If $A$ has group inverse $A$ is regular. From the canonical decomposition of Prasad we can write $A = e_1 A + e_2 A + \cdots + e_k A$ where each $e_i A$ is Rao-regular with Rao-idempotent $e_i$ and $\{e_1, e_2, \cdots, e_k\}$ is a set of pairwise orthogonal idempotents. Since $A$ has group inverse, each $e_i A$ has group inverse.

For any fixed $i$, consider $e_i A$ as a Rao-regular matrix over the Banach algebra $e_i \mathcal{B}$ with $e_i$ as the identity element. From the previous Proposition we have that $\widehat{e_i A}(\phi)$ has group inverse for every $\phi \in \mathcal{M}(e_i \mathcal{B})$.

Now, if $\phi \in \mathcal{M}$, $\hat{A}(\phi) = \sum_i \widehat{e_i A}(\phi_i)$ where $\phi_i(e_i x) = \phi(e_i x)$ for $x \in \mathcal{B}$. Since each $\widehat{e_i A}(\phi_i)$ has group inverse and since $\{e_1, e_2, \cdots, e_k\}$ is a pairwise orthogonal set of idempotents we have that $\hat{A}(\phi)$ has group inverse. Thus (i) $\Rightarrow$ (ii).

To show (ii) $\Rightarrow$ (i), if $\hat{A}(\phi)$ has group inverse for every $\phi \in \mathcal{M}$ then for every idempotent $e, \widehat{eA}(\phi)$ has group inverse for every $\phi \in \mathcal{M}(e\mathcal{B})$. Since $A$ is regular, the canonical decomposition Theorem of Prasad and Theorem 8.20 gives us that $A$ has group inverse.

The equivalence of (ii), (iii) and (iv) is clear using the results of section 7.7.

♠

The results for existence of Drazin inverses can be formulated using the results of section 7.8.

## 8.6 Group inverses in a ring

A major problem which is not touched upon in this book, which was not attempted in the literature also till now, is to characterize regular matrices over an associative ring (possibly noncommutative) with 1. This problem probably needs new techniques.

In this section we shall foray into the uncharted territory of group inverses of elements of an associative ring (possibly noncommutative) with 1. Surprisingly the results turn out to be neat. In the next section we shall use the results of this section to give a complete characterization of elements of an associative ring(possibly noncommutative) with 1 and with an involution $a \to \bar{a}$ admitting Moore-Penrose inverses.

We shall give three approaches to the problem of characterizing elements admitting group inverses.

Let $R$ be an associative ring with 1. Recall that if $a$ and $g$ are elements of $R$ then $g$ is called a group inverse of $a$ if

$$aga = a \tag{1}$$
$$gag = g \tag{2}$$
$$\text{and} \quad ag = ga \tag{5}$$

If $g$ and $a$ satisfy (1) and (5) then $g$ is called a commuting $g$-inverse of $a$. Though commuting $g$-inverse is not unique group inverse when it exists is unique. Clearly, if $g$ is a commuting $g$-inverse of $a$ then $h = gag$ is the group inverse of $a$ and is denoted by $a^{\#}$.

Also, if $g$ is a commuting $g$-inverse of $a$ then $a^2g = a$ and $ga^2 = a$. Conversely, suppose that $a$ and $g$ satisfy $a^2g = a$ and $ga^2 = a$. Then $ga = ga^2g = ag$. Also, $aga = aag = a^2g = a$. Thus, if $a^2g = a$ and $ga^2 = a$ then $g$ is a commuting $g$-inverse of $a$ ($g$ need not be a group inverse).

More generally, if $x$ and $y$ are elements of $R$ such that $a^2x = a$ and $ya^2 = a$ then $g = yax$ is not only a commuting $g$-inverse of $a$ but also is the group inverse of $a$. Indeed, if $a^2x = a$ and $ya^2 = a$ then $ax = ya^2x = ya$. Also, $aga = ayaxa = aaxxa = axa = yaa = a, gag = yaxayax = yyaayax = yaaxx = yax = g$ and $ag = ayax = aaxx = ax = ya = yyaa = yaxa = ga$.

Thus we have proved the following simple Proposition.

**Proposition 8.22:** Let $a$ be an element of an associative ring $R$ with 1. Then $a$ has group inverse if and only if $a^2x = a$ and $ya^2 = a$ both have solutions. In case $a^2x = a$ and $ya^2 = a$ then $a^{\#} = yax$.

♠

We shall now give a more useful characterization.

**Proposition 8.23:** Let $a$ be an element of an associative ring $R$ with 1. Then the following are equivalent.

(i) $a$ has group inverse.

(ii) $a$ is regular and if $a^-$ is a $g$-inverse of $a$ then $a^2a^- - aa^- + 1(= u$, say) is a unit of $R$.

(iii) $a$ is regular and if $a^-$ is a $g$-inverse of $a$ then $a^-a^2 - a^-a + 1(= v$, say) is a unit of $R$.

If $u$ is a unit of $R$ then $a^{\#} = u^{-2}a$. If $v$ is a unit of $R$ then $a^{\#} = av^{-2}$.

**Proof:** We shall show (i) $\Leftrightarrow$ (ii). (i) $\Leftrightarrow$ (iii) can be shown along the same lines.

If $a$ has group inverse $g$ then $a^2g = a$, $ga^2 = a$ and $ag = ga$. Hence, $u(aga^- - aa^- + 1) = 1$ and $(aga^- - aa^- + 1)u = (gaa^- - aa^- + 1)u = 1$. Thus $u$ is a unit of $R$.

To show the converse, let $u^{-1} = \alpha$. Observe that $aa^-u = a^2a^-$ and $ua = a^2$. Hence $a = aa^-a = aa^-u\alpha a = a^2a^-\alpha a$ and also $a = \alpha ua = \alpha a^2$. By the previous Proposition $a^\#$ exists and $a^\# = \alpha aa^-\alpha a$. But, since $aa^-u = a^2a^- = uaa^-$, we have $\alpha aa^- = aa^-\alpha$. Thus $a^\# = \alpha aa^-\alpha a = \alpha^2 a = u^{-2}a$. Similarly $a^\# = av^{-2}$ also follows.

Thus the Proposition is proved.

♠

In the next section we shall apply the results of this section to the problem of finding the group inverse of the companion matrix.

We shall now present a third approach.

**Proposition 8.24:** Let $a$ be an element of an associative ring $R$ with 1. Then $a$ has group inverse if and only if there is an idempotent $p$ such that $a + p$ is a unit and $ap = pa = 0$. Such a $p$, when exists, is unique.

**Proof:** Let $h$ be the group inverse of $a$. Let $p = 1 - ha = 1 - ah$. Clearly, $p$ is idempotent and $ap = pa = 0$. Also, $(a + p)(h + p) = 1 = (h + p)(a + p)$ since $ph = hp = 0$. Thus we have shown that $a + p$ is a unit.

Conversely, if $p$ is an idempotent such that $a+p$ is a unit and $ap = pa = 0$, then, let $g = (a + p)^{-1}(1 - p)$. Note that $(1 - p)(a + p) = a = (a + p)(1 - p)$. Hence, $a(a + p)^{-1} = 1 - p = (a + p)^{-1}a$. Now, let us verify that $g$ is the group inverse of $a$.

$$\begin{aligned} aga &= a(a+p)^{-1}(1-p)a = (1-p)a = a, \\ gag &= (a+p)^{-1}(1-p)a(a+p)^{-1}(1-p) = (a+p)^{-1}(1-p) = g, \\ ag &= a(a+p)^{-1}(1-p) = 1-p \text{ and} \\ ga &= (a+p)^{-1}(1-p)a = 1-p. \end{aligned}$$

Thus, $g$ is the group inverse of $a$. For the uniqueness, observe that if $p$ is an idempotent such that $a + p$ is a unit and $ap = pa = 0$ then it defines a group inverse $g$ of $a$ such that $ag = 1 - p$. From the uniqueness of the group inverse we get the uniqueness of $p$.

♠

## 8.7 M-P inverses in a ring

Let $R$ be an associative ring with 1 and with an involution $a \to \bar{a}$. We shall use the results on existence of group inverses in an associative ring with 1 of the previous section and obtain necessary and sufficient conditions for the existence of Moore-Penrose inverse.

**Theorem 8.25:** Let $a$ be an element of an associative ring $R$ with 1 and with an involution $a \to \bar{a}$. Then the following are equivalent

(i) $a^+$ exists.

(ii) $a^*a$ has group inverse and $a$ satisfies the condition that $ax = 0$ whenever $a^*ax = 0$.

(iii) $aa^*$ has group inverse and $a$ satisfies the condition that $ya = 0$ whenever $yaa^* = 0$.

**Proof:** We shall show (i) $\Leftrightarrow$ (ii). (i)$\Leftrightarrow$ (iii) can be shown along the same lines.

If $a^+$ exists then,

$$a^{+^*}a^*a = (aa^+)^*a = aa^+a = a.$$

Hence $ax = 0$ holds whenever $a^*ax = 0$. That, $a^*a$ has group inverse can be easily shown by showing that $gg^*$ is the group inverse of $a^*a$ if $g$ is the Moore-Penrose inverse of $a$.

Conversely, let $h$ be the group inverse of $a^*a$ and let $a$ satisfy the condition that $ax = 0$ whenever $a^*ax = 0$. Then, $a^*aha^*a = a^*a$ gives us that $a^*a(ha^*a - 1) = 0$. From the property of $a$ that $ax = 0$ whenever $a^*ax = 0$, we get that $aha^*a = a$. Let us show that $g = ha^*$ is the Moore-Penrose inverse of $a$. First note that, since the group inverse is unique and since $(a^*a)^* = a^*a$, we get that $h^* = h$.

Now,

$gag = ha^*aha^* = ha^* = g$ since $ha^*ah = h$.

$(ag)^* = (aha^*)^* = (ah^*a^*)^* = aha^* = ag$, and

$(ga)^* = (ha^*a)^* = a^*ah^* = a^*ah = ha^*a = ga$.

Thus $g$ is the Moore-Penrose inverse of $a$.

♠

From this Theorem it follow that an $m \times n$ matrix $A$ over an associative ring $R$ (possibly noncommutative) with identity and with an involution $a \to \bar{a}$ has Moore-Penrose inverse if and only if there is an idempotent matrix $P$ such that $A^*A + P$ is unimodular, $A^*AP = PA^*A = 0$ and also that $A$ has the property that $AX = 0$ whenever $A^*AX = 0$.

## 8.8 Group inverse of the companion matrix

We shall consider the companion matrix

$$L = \begin{bmatrix} 0 & 0 & \cdots & 0 & a \\ 1 & 0 & \cdots & 0 & b \\ 0 & 1 & \cdots & 0 & a_2 \\ & & \cdots & & \\ 0 & 0 & \cdots & 1 & a_{n-1} \end{bmatrix}$$

over an associative ring(possibly noncommutative) $R$ with 1. Using the results of the previous section we shall find necessary and sufficient conditions for $L$ to have group inverse and find the group inverse when it exists.

**Proposition 8.26:** Let $L$ be the companion matrix over a general ring $R$ with 1. Then the following are equivalent.

(i) $L^{\#}$ exists.

(ii) $a$ is regular and $a + (1 - aa^-)b$ is a unit.

(iii) $a$ is regular and $a + b(1 - a^-a)$ is a unit.

(iv) $a$ is regular and $a - (1 - aa^-)b$ is a unit.

(v) $a$ is regular and $a - b(1 - a^-a)$ is a unit.

**Proof:** By Proposition 8.23 $L$ has group inverse if and only if $L$ is regular and $U = L^2L^- - LL^- + I$ where $L^-$ is a $g$-inverse of $L$ is a unit. Hence $a$ must be regular. Let $a^-$ be a $g$-inverse of $a$.

We shall first consider the case $n > 2$.

An obvious

$$L^- = \begin{bmatrix} -ba^- & 1 & 0 & \cdots & 0 \\ -a_2a^- & 0 & 1 & \cdots & 0 \\ -a_3a^- & 0 & 0 & \cdots & 0 \\ . & . & . & \cdots & . \\ -a_{n-1}a^- & 0 & 0 & \cdots & 1 \\ a^- & 0 & 0 & \cdots & 0 \end{bmatrix}.$$

Hence

$$LL^- = \begin{bmatrix} aa^- & 0 & 0 & \cdots & 0 \\ 0 & 1 & 0 & \cdots & 0 \\ . & . & . & \cdots & . \\ 0 & 0 & 0 & \cdots & 1 \end{bmatrix}$$

$$L^2L^- = \begin{bmatrix} 0 & 0 & \cdots & 0 & a \\ aa^- & 0 & \cdots & 0 & b \\ 0 & 1 & \cdots & 0 & a_2 \\ . & . & \cdots & . & . \\ 0 & 0 & \cdots & 1 & a_{n-1} \end{bmatrix}$$

and

$$U = L^2L^- - LL^- + I = \begin{bmatrix} 1-aa^- & 0 & 0 & \cdots & 0 & a \\ aa^- & 0 & 0 & \cdots & 0 & b \\ 0 & 1 & 0 & \cdots & 0 & a_2 \\ 0 & 0 & 1 & \cdots & 0 & a_3 \\ . & . & . & \cdots & . & . \\ 0 & 0 & 0 & \cdots & 1 & a_{n-1} \end{bmatrix}.$$

Now, $U$ is a unit if and only if

$$\begin{bmatrix} 1-aa^- & a \\ aa^- & b \end{bmatrix}$$

is a unit. $U^{-1}$ can be constructed from the inverse of

$$\begin{bmatrix} 1-aa^- & a \\ aa^- & b \end{bmatrix}$$

by using the formula

$$\begin{bmatrix} B & 0 \\ C & I \end{bmatrix}^{-1} = \begin{bmatrix} B^{-1} & 0 \\ -CB^{-1} & I \end{bmatrix}.$$

Let $e = 1 - aa^-$. Then $e$ is an idempotent, $1-e$ is an idempotent, $ea = 0$ and $(1-e)a = a$.

So we shall find necessary and sufficient conditions for the existence of the inverse of

$$\begin{bmatrix} e & a \\ 1-e & b \end{bmatrix} = C \text{ (say)}.$$

Let $D = \begin{bmatrix} e & 1-e \\ 1-e & e \end{bmatrix}$. Then $DD = I$ or $D^{-1} = D$.

Let $E = DC = \begin{bmatrix} e & 1-e \\ 1-e & e \end{bmatrix}\begin{bmatrix} e & a \\ 1-e & b \end{bmatrix} = \begin{bmatrix} 1 & (1-e)b \\ 0 & a+eb \end{bmatrix}$.

Hence $C$ is a unit if and only if $a+eb$ is a unit. Thus $L$ has group inverse if and only if $a$ is regular and $a + (1-aa^-)b$ is a unit. Thus (i) $\Leftrightarrow$ (ii) for the case of $n > 2$ is proved. We shall postpone the case of $n = 2$ to the end.

We shall proceed towards finding the exact expression for the $L^{\#}$.

Now, $E^{-1} = \begin{bmatrix} 1 & -(1-e)b(a+eb)^{-1} \\ 0 & (a+eb)^{-1} \end{bmatrix}$

and
$$C^{-1} = E^{-1}D = \begin{bmatrix} 1 & -(1-e)b(a+eb)^{-1} \\ 0 & (a+eb)^{-1} \end{bmatrix}\begin{bmatrix} e & 1-e \\ 1-e & e \end{bmatrix}$$
$$= \begin{bmatrix} e-(1-e)b(a+eb)^{-1}(1-e) & 1-e-(1-e)b(a+eb)^{-1}e \\ (a+eb)^{-1}(1-e) & (a+eb)^{-1}e \end{bmatrix}$$

Let this matrix be $\begin{bmatrix} s & t \\ u & v \end{bmatrix}$. This is $\begin{bmatrix} e & a \\ 1-e & b \end{bmatrix}^{-1}$.

Now

$$U^{-1} = \begin{bmatrix} s & t & 0 & \cdots & 0 \\ -a_2u & -a_2v & 1 & \cdots & 0 \\ . & . & . & \cdots & . \\ -a_{n-1}u & -a_{n-1}v & 0 & \cdots & 1 \\ u & v & 0 & \cdots & 0 \end{bmatrix}$$

Since $L^{\#} = U^{-2}L$, we find $U^{-1}L$ first.

$$U^{-1}L = \begin{bmatrix} s & t & 0 & \cdots & 0 \\ -a_2u & -a_2v & 1 & \cdots & 0 \\ . & . & . & \cdots & . \\ -a_{n-1}u & -a_{n-1}v & 0 & \cdots & 1 \\ u & v & 0 & \cdots & 0 \end{bmatrix} \begin{bmatrix} 0 & 0 & \cdots & 0 & a \\ 1 & 0 & \cdots & 0 & b \\ 0 & 1 & \cdots & 0 & a_2 \\ . & . & \cdots & . & . \\ 0 & 0 & \cdots & 1 & a_{n-1} \end{bmatrix}$$

$$= \begin{bmatrix} t & 0 & 0 & \cdots & 0 & sa+tb \\ -a_2v & 1 & 0 & \cdots & 0 & -a_2ua - a_2vb + a_2 \\ -a_3v & 0 & 1 & \cdots & 0 & -a_3ua - a_3vb + a_3 \\ . & . & . & \cdots & . & . \\ -a_{n-1}v & 0 & 0 & \cdots & 1 & -a_{n-1}ua - a_{n-1}vb + a_{n-1} \\ v & 0 & 0 & \cdots & 0 & ua+vb \end{bmatrix}$$

But, since $sa + tb = 0$ and $ua + vb = 1$ we get that

$$U^{-1}L = \begin{bmatrix} t & 0 & 0 & \cdots & 0 & 0 \\ -a_2v & 1 & 0 & \cdots & 0 & 0 \\ -a_3v & 0 & 1 & \cdots & 0 & 0 \\ . & . & . & \cdots & . & . \\ -a_{n-1}v & 0 & 0 & \cdots & 1 & 0 \\ v & 0 & 0 & \cdots & 0 & 1 \end{bmatrix}$$

$$\begin{aligned} L^{\#} &= U^{-2}L \\ &= U^{-1}U^{-1}L \\ &= \begin{bmatrix} s & t & 0 & \cdots & 0 \\ -a_2u & -a_2v & 1 & \cdots & 0 \\ . & . & . & \cdots & . \\ -a_{n-1}u & -a_{n-1}v & 0 & \cdots & 1 \\ u & v & 0 & \cdots & 0 \end{bmatrix} \begin{bmatrix} t & 0 & 0 & \cdots & 0 \\ -a_2v & 1 & 0 & \cdots & 0 \\ -a_3v & 0 & 1 & \cdots & 0 \\ . & . & . & \cdots & . \\ -a_{n-1}v & 0 & 0 & \cdots & 1 \\ v & 0 & 0 & \cdots & 0 \end{bmatrix} \end{aligned}$$

$$= \begin{bmatrix} st - ta_2v & t & 0 & 0 & \cdots & 0 \\ -a_2ut + a_2va_2v - a_3v & -a_2v & 1 & 0 & \cdots & 0 \\ -a_3ut + a_3va_2v - a_4v & -a_3v & 0 & 1 & \cdots & 0 \\ . & . & . & . & \cdots & . \\ -a_{n-1}ut + a_{n-1}va_2v + v & -a_{n-1}v & 0 & 0 & \cdots & 1 \\ ut - va_2v & v & 0 & 0 & \cdots & 0 \end{bmatrix}.$$

For the case $n = 2$

$$L = \begin{bmatrix} 0 & a \\ 1 & b \end{bmatrix}$$

and if $L^{\#}$ exists clearly $a$ must be regular. If $a^-$ is a $g$-inverse of $a$ then, as before,

$$U = \begin{bmatrix} 1 - aa^- & a \\ aa^- & b \end{bmatrix} = \begin{bmatrix} e & a \\ 1 - e & b \end{bmatrix}$$

where $e = 1 - aa^-$. For $L^{\#}$ to exist $U$ must be unimodular. As before,

$$U^{-1} = \begin{bmatrix} s & t \\ u & v \end{bmatrix} \text{ exists}$$

if and only if $a + (1 - aa^-)b$ is a unit. In this case,

$$L^{\#} = U^{-2}L = \begin{bmatrix} st + tv & t \\ ut + v^2 & v \end{bmatrix}.$$

Thus (i) $\Leftrightarrow$ (ii) is proved.

(i) $\Leftrightarrow$ (iii) also follows in a similar way considering the matrix $L^-L^2 - L^-L + I$.

Note that $(1 - 2e)(a + eb) = a - eb$ and $1 - 2e$ is a unit. Thus the rest of the implications are clear.

♠

# EXERCISES:

**Exercise 8.1:** Show that for a regular matrix $A$ existence of bordering neither implies nor is implied by the existence of rank factorization for $A$. (See the example on p. 254 of [72]).

**Exercise 8.2:** Let $R$ be a commutative ring. Let $L$ be the companion matrix as above. Using the decomposition Theorem of Robinson for $L$ as in section 7.4 and the conditions for the existence of Moore-Penrose inverse and the Group inverse as in sections 7.6 and 7.7 find necessary and sufficient conditions on the elements of $L$ so that $L$ admits Moore-Penrose inverse (and the Group inverse). Also find the Moore-Penrose inverse (and the Group inverse of $L$) when they exist using this method.

**Exercise 8.3:** If $R$ is a commutative ring with the property that for every finitely generated subring $R'$ of $R$ with identity there is a projective free subring $R''$ such that $R' \subset R'' \subset R$ then show that $R$ is projective free (Hint: use the equivalence (i) $\Leftrightarrow$ (ii) of Theorem 8.15).

**Exercise 8.4:** We have shown earlier that if $G$ is a reflexive $g$-inverse of a Rao-regular matrix $A$ over a commutative ring $R$ with $\rho(A) = r$ then for $(\alpha, \beta) \in \Delta_{k,n,m}$

$$|G_{\alpha\beta}| = \sum_{(\gamma,\delta)\in\Delta_{r,m,n}} |G_{\delta\gamma}| \frac{\partial |A_{\gamma\delta}|}{\partial |A_{\beta\alpha}|}.$$

Using the results of this Chapter show that such a formula holds good even when $G$ is a reflexive $g$-inverse of a general matrix $A$ with the property that $I - AG$ and $I - GA$ both have rank factorizations. (Here $R$ need not be projective free and $A$ need not be Rao-regular).

**Exercise 8.5:** Similar to the results of section 8.6, find necessary and sufficient conditions for the existence of Drazin inverse of an element $a$ of an associative ring $R$ with 1.

**Exercise 8.6:** Let $R$ be an associative ring with 1 and with an involution $a \to \bar{a}$. Let $a$ be an element of $R$. Show that $a$ has a $\{1,3\}$-inverse if and only if $a^*a$ is regular and $a$ has the property that $ax = 0$ whenever $a^*ax = 0$. Compare with Proposition 3.10. Also obtain similar results for the other types of $g$-inverses.

**Exercise 8.7:** If an element $a$ in an associative ring with 1 has group inverse find the relation of $u$ of Proposition 8.23 to $p$ of Proposition 8.24?

# Bibliography

[1] Ballico, E., Rank factorization and Bordering of regular matrices over commutative rings, *Linear Algebra and Its Applications,* 305(2000), 187–190.

Borderings of different sizes and their relation to rank factorizations are considered. This paper is related to [72] by Prasad and Bhaskara Rao.

[2] Bapat, R.B. Bhaskara Rao, K.P.S. and Prasad, K.M., Generalized inverses over integral domains, *Linear Algebra and Its Applications,* 140(1990), 180–196.

Some of the results of Chapter 6 are taken from here.

[3] Bapat, R.B. and Robinson D.W., The Moore-Penrose inverse over a commutative ring, *Linear Algebra and Its Applications,* 177(1992), 89–103.

Necessary and sufficient conditions for a matrix over a commutative ring to admit Moore-Penrose inverse are obtained. This paper is a forerunner to the paper of Prasad [71] in which a complete characterization of regularity of matrices over commutative ring is obtained.

[4] Bapat, R.B., Generalized inverses with proportional minors, *Linear Algebra and Its Applications,* 211(1994), 27–33.

[5] Bapat, R.B. and Ben-Israel, A., Singular values and maximum rank minors of Generalized inverses, *Linear and Multilinear Algebra,* 40(1995), 153–161.

Theorem 8.7 says that for a Rao-regular matrix $A$ of rank $r$ when $A^+$ exists, $A^+$ and $A^*$ have the same $r \times r$ minors except for a proportionality constant. Theorem 8.8 tells us a similar result for group inverses. The concept of proportionality of minors was studied in the above two papers. If $A$ and $H$ are matrices of rank $r$ over an integral domain, it is shown that $A$ admits a $g$-inverse whose $r \times r$ minors are proportional to the $r \times r$ minors of $H$ if and only if $Tr(C_r(AH))$ is a unit. The relation to the volumes of matrices are also studied.

[6] Barnett, C. and Camillo. V., Idempotents in Matrices over commutative Von Neumann regular rings, *Comm. Algebra,* 18(1990), 3905–3911.

[7] Barnett, C. and Camillo. V., Idempotents in Matrix Rings, *Proc. Amer. Math. Soc.,* 122(1994), 965–969.

The above two papers deal with characterizing idempotent matrices over Von Neumann regular rings. Theorem 3 of the second paper gives a decomposition Theorem for idempotent matrices. This is same as the decomposition Theorem of Prasad for idempotent matrices.

[8] Barnett, S., *Matrices in Control Theory,* Van Nostrand (1971).

[9] Batigne, D.R., Integral generalized inverses of integral matrices, *Linear Algebra and Its Applications,* 22(1978), 125–135.

This is the first paper on Moore-Penrose inverses of matrices with integer entries. The results were obtained without the use of Smith Normal Form Theorem. For regularity of matrices over integers, more generally over a principal ideal domain, Smith Normal Norm Theorem was found to be useful in the paper of Bhaskara Rao [17] and in that of Bose and Mitra [22].

[10] Batigne, D.R., Hall, F.J. and Katz, I.J., Further results on integral generalized inverses of integral matrices, *Linear and Multilinear Algebra,* 6(1978), 233–241.

For matrices over $\mathbb{Z}$ the existence of $\{1,3\}$-inverses and $\{1,4\}$-inverses are studied. The relations to solutions of linear equations are also studied.

[11] Beasley, Leroy B. and Pullman, Norman J., Semiring Rank versus Column Rank, *Linear Algebra and Its Applications,* 101(1988), 33–48.

Two of the notions of rank of matrices over a semiring, namely, the semiring rank and the column rank ($S(A)$ and $c(A)$ as in section 2.4 in the present monograph) are compared. These two notions are the same for the semirings of fields and Euclidean domains. But they differ over other algebraic structures.

[12] Ben-Israel, A. and Greville, T.N., *Generalized Inverses: Theory and Applications,* John Wiley & Sons, Inc. New York, 1974.

This is a rich source of information on the classical theory of Generalized inverses.

[13] Ben-Israel, A., A Cramer's Rule for Least-Square Solution of Consistent Linear Equations, *Linear Algebra and Its Applications,* 43(1982), 223–228.

Some of the treatment of section 8.1 is based on this paper.

[14] Ben-Israel, A., A volume associated with $m \times n$ matrices, *Linear Algebra and Its Applications,* 167(1992), 87–111.

The volume of an $m \times n$ real matrix $A$ of rank $r$ is defined as $\sqrt{\sum |A_{\alpha,\beta}|^2}$. The relations between the volume of a matrix and the Moore-Penrose inverse are investigated in this paper.

[15] Berenstein, C.A. and Struppa, D.C., On Explicit Solution to the Bezout Equation, *System and Control Letters,* 4(1984), 33–39.

[16] Bhaskara Rao, K.P.S. and Prasada Rao, P.S.S.N.V., On generalized inverses of Boolean matrices, *Linear Algebra and Its Applications,* 11(1975), 135–153.

See the annotation for [18].

[17] Bhaskara Rao, K.P.S., On generalized Inverses of Matrices over principal ideal domains. *Linear and Multilinear Algebra,* 10(1980), 145–154.

The Smith normal form Theorem was used to determine the regular matrices over principal ideal domains. This paper has some overlap with the paper of Bose and Mitra [22].

[18] Bhaskara Rao, K.P.S., On generalized Inverses of Matrices over principal ideal domains-II, (1981), *unpublished.*

Moore-Penrose inverses and $\{1,3\}$-inverses and $\{1,4\}$-inverses and their significance in the context of polynomial rings over fields are investigated in this unpublished paper.

[19] Bhaskara Rao, K.P.S. and Prasada Rao, P.S.S.N.V., On generalized inverses of Boolean matrices. II, *Linear Algebra and Its Applications,* 42(1982), 133–144.

Papers [16] and [19] give a complete treatment of regularity of matrices over the $\{0, 1\}$ Boolean algebra. A technique for treating matrices over the general Boolean algebra was also explained. Most of the results of the first paper are reported in the Book of Kim [53].

[20] Bhaskara Rao, K.P.S., On generalized Inverses of Matrices over Integral Domains, *Linear Algebra and Its Applications,* 49(1983), 179–189.

The first ever result characterizing the existence of $g$-inverses over integral domains in terms of minors was proved here.

[21] Bhaskara Rao, K.P.S., Generalized Inverses of Matrices over rings, (1985), *unpublished.*

The Basic Theorem 5.3 appeared for the first time in this paper. Also regular matrices over the ring $\mathbb{Z}[k]$ and also over the ring of real valued continuous functions on the real line (Examples 1 and 2 of section 7.1) were characterized in this paper.

[22] Bose, N.K. and Mitra, S.K., Generalized inverses of polynomial matrices, *IEEE Trans. Automatic Control,* AC-23(1978), 491–493.

Smith Normal form Theorem was used for investigating the regularity of polynomial matrices.

[23] Bourbaki, N., *Commutative Algebra,* Addison Wesley, 1972.

[24] Brown, B. and McCoy, N. H., The maximal regular ideal of a ring, *Proc. Amer. Math. Soc.,* 1(1950), 165–171.

The proof of the Theorem of Von Neumann in section 3.3 is taken from here.

[25] Brown, William C., *Matrices over commutative rings,* Marcel Dekker, 1993.

[26] Bruening, J.T., A new formula for the Moore-Penrose inverse, *Current Trends in Matrix Theory,* F. Uhlig and R. Grone, Eds., Elsevier science, 1987.

For full row rank matrices $A$, a formula for the Moore-Penrose inverse using minors of $A$ was obtained here.

[27] Cao, Chong-Guang, Generalized inverses of matrices over rings (in Chinese), *Acta Mathematica Sinica,* 31(1988), 131–133.

Some of the results of section 3.5 are based on this paper.

[28] Chen, Xuzhou and Hartwig, R.E., The Group Inverse of a Triangular Matrix, *Linear Algebra and Its Applications,* 237/238(1996), 97–108.

Simple block conditions for the existence of a group inverse of a triangular matrix over a field are given.

[29] Cho, Han H., Regular Matrices in the Semigroup of Hall Matrices, *Linear Algebra and Its Applications,* 191(1993), 151–163.

Consider the Boolean algebra $\mathcal{B}$ of two elements 0 and 1. For a positive integer $s$ let $H_n(s)$ be the semigroup of all $n \times n$ matrices $A$ over $\mathcal{B}$ such that the permanent of $A$ is $\geq s$. Regularity of matrices in $H_n(s)$ is characterised. Any matrix in any of the $H_n(s)$ for a positive integer $s$ is called a *Hall matrix.* Many interesting results on regularity of Hall matrices were obtained.

[30] Deutsch, E., Semi Inverses, reflexive Semi Inverses and Pseudo Inverses of an Arbitrary Linear Transformation, *Linear Algebra and Its Applications,* 4(1971), 95–100.

[31] Drazin, M.P., Pseudo-inverses in associative rings and semigroups, *Amer. Math. Monthly,* 65(1958), 506–514.

Drazin inverse was introduced in this paper. Its uniqueness and many other basic results were proved in this paper.

[32] Elizanov, V.P., Systems of Linear equations over a commutative ring, *Uspehi Math. Nauk.* 2(290), 181–182; *Russian Mathematical Surveys,* 48(1993) No.2, 175–177.

If $R$ is a chain ring (= commutative local ring), for a matrix $A$ over $R$ it is shown that $A\underline{x} = \underline{b}$ has a solution if and only if $\mathcal{I}_m(A) = \mathcal{I}_m([A \;\; \underline{b}])$.

[33] Engelking, R., *Outline of General Topology,* North Holland, 1968.

[34] Fulton, J.D., Generalized Inverse of matrices over finite field, *Discrete Mathematics,* 21(1978), 23–29.

The number of various types of $g$-inverses of a given matrix over a finite field is counted. Also the number of matrices which admit $\{1,2,3\}$-inverses and $\{1,2,4\}$-inverses are also counted.

[35] Gelfand, I., Raikov, D. and Shilov, G., *Commutative normed rings,* Chelsea, 1964.

[36] Goodearl, K.R., *Von Neumann regular rings,* Pitman, New York, 1978.

[37] Gouveia, M.C. and Puystjens, R., About the group inverse and Moore-Penrose inverse of a product, *Linear Algebra and Its Applications,* 150(1991), 361–369.

Necessary and sufficient conditions are obtained for a matrix $PAQ$ to admit a group inverse (Moore-Penrose Inverse) if $A$ has group inverse (has Moore-Penrose inverse) under some conditions on $P$ and $Q$.

[38] Gouveia, M.C., Generalized invertibility of Hankel and Toeplitz matrices *Linear Algebra and Its Applications,* 193(1993), 95–106.

See the annotation for [40].

[39] Gouveia, M.C., Group and Moore-Penrose invertibility of Bezoutians, *Linear Algebra and Its Applications,* 197(1993), 495–509.

Generalized inverses of Bezoutians, generalized Bezoutians and related matrices are investigated.

[40] Gouveia, M.C., Regular Hankel Matrices Over Integral Domains, *Linear Algebra and Its Applications,* 255(1997), 335–347.

An $m \times n$ matrix $H = (h_{ij})$ for $0 \leq m-1, 0 \leq n-1$ where $h_{ij} = a_{i+j}$ where $a_0, a_1, \cdots, a_{m+m-2}$ are elements of a ring is called a Hankel matrix. Various aspects of regularity of Hankel matrices over fields and integral domains are studied in [38] and [40].

[41] Gregory, D.A. and Pullman, Norman J., Semiring Rank: Boolean Rank and nonnegative rank factorizations, *J. Combin. Inform. System. Sci.,* 8(1983), 223–233.

Some results on ranks of matrices over some algebraic structures are studied.

[42] Hansell, G.W. and Robinson, D.W., On the Existence of Generalized Inverses, *Linear Algebra and Its Applications,* 8(1974), 95–104.

[43] Harte, R. and Mostafa, M., On generalized inverses in $C^*$-algebras, *Studia Mathematica,* 103(1992), 71–77.

[44] Harte, R. and Mostafa, M., Generalized inverses in $C^*$-algebras II, *Studia Mathematica,* 106(1993), 129–138.

Generalized inverses of elements of a $C^*$-algebra are studied in the above two papers. For example, it is shown that if an element of a $C^*$-algebra is regular then it has Moore-Penrose inverse. Many other results are obtained.

[45] Henriksen, M., personal communication, 1989.

Some of the results on integral domains satisfying the Rao condition are taken from here.

[46] Huang, D.R., Generalized inverses over Banach Algebras, *Integral Equations Operator Theory,* 15(1992), 454–469.

A Banach algebra is a nice commutative ring having zero divisors. Using the theory of Banach algebras regular matrices over Banach algebras are characterized. This is a forerunner to the results of Bapat and Robinson [3].

[47] Huang, D.R., Generalized inverses and Drazin inverses over Banach algebras, *Integral Equations Operator Theory,* 17(1993).

Group and Drazin inverses of matrices over a Banach algebra are considered. Many of the results of the above two papers are included in the present book.

[48] Huylebrouck, D. Puystjens, R. and Van Geel, J., The Moore-Penrose Inverse of a matrix over a semisimple Artinian Ring, *Linear and Multilinear Algebra,* 16(1984), 239–246.

Necessary and sufficient conditions for the existence of the Moore-Penrose inverse of a matrix over a semi-simple Artinian ring with an involution are obtained.

[49] Huylebrouck, D. and Puystjens, R., Generalized inverses of um with a radical element, *Linear Algebra and Its Applications,* 84(1986), 289–300.

[50] Huylebrouck, D., Diagonal and Von Neumann regular Matrices over a Dedekind domain, *Portugalae Mathematicae,* 51(1994), No.2, 291–293.

[51] Jacobson, N., *Basic Algebra,* I, II, III, Freeman and Company, (1967–1980).

[52] Karampetakis, N.P., Computation of the Generalized Inverse of a Polynomial Matrix and Applications, *Linear Algebra and Its Applications,* 252(1997), 35–60.

An algorithm for computing the Moore-Penrose inverse of a matrix over the ring of polynomials with real coefficients is given. Applications to control theory are also explained. This paper has many references to applications.

[53] Kim, K.H., *Boolean Matrix theory and Applications,* Pure and Applied Mathematics, Vol. 70, Marcel Dekker, New York, 1982.

This is a treatise on Boolean Matrices which also includes a good treatment of regular Boolean matrices.

[54] Koliha, J.J. and Rakocevic, V., Continuity of the Drazin inverse II, *Studia Mathematica,* 131(1998), 167–177.

Continuity of the generalized Drazin inverse for elements of a Banach algebra and bounded linear operators on Banach spaces is studied.

[55] Lam, T.Y., *Serre's conjecture,* Lecture notes in mathematics, V.635 Springer-Verlag(1978).

[56] Lovass-Nagy, V., Miller, R.J. and Powers, D.L., An introduction to the application of the simplest matrix generalized inverse in systems, *IEEE Transactions on circuits and systems science,* 25(1978), 766–771.

[57] McCoy, N.H., *Rings and Ideals,* Carus Math. Monograph No.8, Math. Association of America, Washington, D.C., 1948.

[58] McDonald, B.R., *Linear algebra over commutative rings,* Marcel Dekker, 1984.

This is a comprehensive and excellent treatise on the theory of matrices over commutative rings.

[59] Miao, J. and Robinson, D.W., Group and Moore-Penrose Inverses of Regular Morphisms with Kernel and Cokernel, *Linear Algebra and Its Applications,* 110(1988), 263–270.

[60] Miao, J., and Ben-Israel, A., Minors of the Moore-Penrose Inverse, *Linear Algebra and Its Applications,* 195(1993), 191–208.

The results of section 8.3 on Jacobi's identity are fashioned after the results of this paper.

[61] Miao, J., Reflexive generalized inverses and their minors, *Linear and Multilinear Algebra,* 35(1993), 153–163.

Generalizes the results of [60] to reflexive $g$-inverses. Some of the results of this paper can be found in section 8.3.

[62] Muir, T., *A treatise on the theory of determinants,* Dover, 1960.

An extremely useful treatise every matrix algebraist should own and look at once in a while.

[63] Nashed, M.Z., *Generalized Inverses and Its Applications,* Academic Press, New York, (1976).

This is a treatise which also has an extensive bibliography upto 1975.

[64] Newman, M., *Integral Matrices,* Academic Press, 1972.

A wonderful book that gives a comprehensive treatment of matrices over principal ideal rings.

[65] Nomakuchi, K., On the characterization of Generalized Inverses by Bordered Matrices, *Linear Algebra and Its Applications,* 33(1980), 1–8.

Bordering for real matrices was considered here. Actually Bordering has a history as given in the book of Ben-Israel and Greville. The results of section 8.4 are generalizations of results of this paper.

[66] Pearl, M.H., Generalized Inverses of Matrices with entries taken from arbitrary field, *Linear Algebra and Its Applications,* 1(1969), 571–587.

Basic results regarding the existence of $\{1, 3\}$-inverses, $\{1, 4\}$-inverses and Moore-Penrose inverses for matrices over fields are obtained here.

[67] Pierce, R.S., *Modules over commutative Von Neumann regular rings,* Mem. Amer. Math. Soc., 70(1967).

[68] Prasad, K.M., Bhaskara Rao, K.P.S. and Bapat, R.B., Generalized inverses over integral domains II: Group and Drazin inverses, *Linear Algebra and Its Applications,* 146(1991), 31–47.

Some of the results of Chapter 6 are taken from here.

[69] Prasad, K.M. and Bapat, R.B., The generalized Moore-Penrose inverse, *Linear Algebra and Its Applications,* 165(1992), 59–69.

The concept of Generalized Moore-Penrose inverse is studied here.

[70] Prasad, K.M. and Bapat, R.B., A note on the Khatri inverse, *Sankhya(Series A): Indian J. of Statistics,* 54(1992), 291–295.

A mistake in a paper of Khatri on a type of $g$-inverse was pointed out in this paper. The correct formulation was also obtained.

[71] Prasad, K.M., Generalized inverses of matrices over commutative rings, *Linear Algebra and Its Applications,* 211(1994), 35–52.

The problem of characterizing regular matrices over commutative rings is solved. The decomposition Theorem for regular matrices in terms of Rao-regular matrices is discovered in this paper.

[72] Prasad, K.M. and Bhaskara Rao, K.P.S., On bordering of regular matrices, *Linear Algebra and Its Applications,* 234(1996), 245–259.

Necessary and sufficient conditions are given for a commutative ring to have the property that every regular matrix can be completed to an invertible matrix of particular size. Such rings are precisely those over which every finitely generated projective module is free. Some of the results of this paper are generalizations of results from the paper of Nomakuchi [65].

[73] Prasad, K.M., Solvability of Linear Equations and rank-function, *Communications in Algebra,* 25(1)(1997), 297–302.

For a regular matrix $A$ over a commutative ring necessary and sufficient conditions are given for the consistency of $A\underline{x} = \underline{a}$ in terms of determinantal rank. Some of the results of section 8.2 are taken from here.

[74] Puystjens, R. and De Smet, H., The Moore-Penrose inverse for matrices over skew polynomial rings, Ring Theory, Antwerp 1980 (Proc. Conf., Univ. Antwerp, Antwerp, 1980), 94–103, *Lecture Notes in Mathematics,* 825, Springer, Berlin, 1980.

[75] Puystjens, R. and Constales, D., The group and Moore-Penrose inverse of companion matrices over arbitrary commutative rings, preprint, University of Gent, Belgium.

[76] Puystjens, R. and Robinson, D.W., The Moore-Penrose inverse of a morphism with a factorization, *Linear Algebra and Its Applications,* 40(1981), 129–141.

[77] Puystjens, R. and Van Geel, J.V., Diagonalization of matrices over graded Principal Ideal Domains, *Linear Algebra and Its Applications,* 48(1982), 265–281.

The theory of graded rings is well-known to algebraists. Similar to theory of diagonalization of matrices over principal ideal domains (the Smith normal form Theorem), a diagonilization result was obtained for graded matrices over graded principal ideal domains. Using this a result was obtained for the existence of a graded $g$-inverse for a graded matrix over a graded principal ideal domain.

[78] Puystjens, R., Moore-Penrose inverses for matrices over some Noetherian rings, *Journal of Pure and Applied Algebra,* 31(1984), 191–198.

Moore-Penrose inverses of matrices over left and right principal ideal domains are studied. Results on $g$-inverses of morphisms in additive categories are used in proving the results.

[79] Puystjens, R. and Robinson, D.W., The Moore-Penrose inverse of a morphism in an additive category, *Communications in Algebra,* 12(1984), 287–299.

[80] Puystjens, R., Some aspects of generalized invertibility, *Bull. Soc. Math. Belg.,* Ser A 40(1988), 67–72.

[81] Puystjens, R. and Robinson, D.W., Symmetric morphisms and the existence of Moore-Penrose inverses, *Linear Algebra and Its Applications,* 131(1990), 51–69.

The three papers [76], [79] and [81] deal with $g$-inverses of morphisms in an additive category. The authors develop a sound theory and obtain several interesting results.

[82] Puystjens, R. and Hartwig, R.E., The Group inverse of a companion matrix, *Linear and Multilinear Algebra,* 43(1997), 135–150.

This is a sequel to [75]. The results of section 8.7 are fashioned after this paper.

[83] Rao, C.R. and Mitra, S.K., *Generalized Inverse of Matrices and Its Applications,* Wiley & Sons, Inc. New York, 1971.

A wealth of information of $g$-inverses of real matrices and their applications is in this book.

[84] Robinson, D.W. and Puystjens, R., EP-morphisms, *Linear Algebra and Its Applications,* 64(1985), 157–174.

[85] Robinson, D.W. and Puystjens, R., Generalized inverses of morphisms with kernels, *Linear Algebra and Its Applications,* 96(1987), 65–86.

Some results on $g$-inverses of morphisms in an additive category are obtained.

[86] Robinson, D.W., Puystjens, R. and Van Geel, J., Categories of matrices with only obvious Moore-Penrose inverses, *Linear Algebra and Its Applications,* 97(1987), 93–102.

The term "Rao condition" was introduced and studied in detail here. Some of the results of section 4.7 are taken from here.

[87] Robinson, D.W., The Moore idempotents of a Matrix, *Linear Algebra and Its Applications,* 211(1994), 15–26.

For a matrix $A$ over a commutative ring with involution $a \to \bar{a}$ and with 1, if $A$ admits Moore-Penrose inverse, the concept of "list of Rao-idempotents" is called the "list of Moore idempotents". The concept of Moore idempotents was introduced before the concept of Rao-idempotents was introduced.

[88] Robinson, D.W., The determinental rank idempotents of a Matrix, *Linear Algebra and Its Applications,* 237/238(1996), 83–96.

Robinson's decomposition Theorem (section 7.4) is proved in this paper. Also the concepts of Rao-regular matrices, Rao-list of idempotents, Rao-list of ranks and Rao index are introduced here.

[89] Robinson, D.W., The image of the adjoint mapping, *Linear Algebra and Its Applications,* 277(1998), 142–148.

The map $S \to A_S$ as in Chapter 5 is called the adjoint mapping in this paper. The image of this mapping is characterized in this paper. In particular, it follows that every $g$-inverse of a matrix $A$ over an integral domain is of the form $A_S$ for some $S$. See Theorem 7.5.

[90] Robinson, D.W., Outer inverses of matrices, preprint.

Consider matrices $A$ and $G$ over a commutative ring with 1. Let $\rho(A) = \rho(G) = r$. and let $Tr(C_r(GA)) = 1$. Then it is shown that $G$ is an outer inverse of $A$ (i. e., 2-inverse of $A$) if and only if $G = A_{C_r(G)}$. Thus 2-inverses of matrices over commutative rings with 1 are characterized in this paper.

[91] Roch, S. and Silberman, B., Continuity of generalized inverses in Banach Algebras, *Studia Mathematica,* 136(1999), 197–227.

The results of Proposition 8.24 and Theorem 8.25 are taken from this paper. The main results of this paper are more functional analytic.

[92] Song, G. and Guo, X., Diagonability of idempotent matrices over noncommutative rings, *Linear Algebra and Its Applications,* 297(1999), 1–7.

Over any associative ring with 1, if $A$ is an idempotent matrix for which there are unimodular matrices $S$ and $T$ such that $SAT$ is a diagonal matrix then there is a unimodular matrix $U$ such that $UAU^{-1}$ is a diagonal matrix. This result generalizes the result of [96] to noncommutative rings.

[93] Sontag, E.D., Linear systems over commutative rings: A survey, *Richerche di Automata,* 7(1976), 1–34.

[94] Sontag, E.D., On generalized inverses of polynomial and other matrices, *IEEE Trans. Automatic Control* AC-25(1980), 514–517.

This is one of the first papers to characterize regular matrices over the ring of polynomials.

[95] Sontag, E.D. and Wang, Y., Pole shifting for families of linear systems depending on at most three parameters, *Linear Algebra and Its Applications,* 137(1990), 3–38.

[96] Steger, A., Diagonability of idempotent matrices, *Pacific Journal of Mathematics,* 19(1996), 535–541.

For an idempotent matrix $A$ over a commutative ring $R$ with 1 if there are unimodular matrices $S$ and $T$ such that $SAT$ is a diagonal matrix then there is a unimodular matrix $U$ such that $UAU^{-1}$ is a diagonal matrix. This result is proved in this paper. See Theorem 8.15 for related results.

[97] Tartaglia, M. and Aversa, V., On the existence of $g$-inverse of a matrix, unpublished, (1999).

Using Gröbner Bases and *Mathematica* the authors provide an algorithm to find a $g$-inverse of any matrix over the ring of polynomials in several variables with real coefficients. They also discuss algorithms for other $g$-inverses.

[98] Von Neumann, J., On regular rings, *Proc. Nat. Acad. Sci.(U S A),* 22(1936), 707–713.

[99] Von Neumann, J., *Continuous Geometry,* Princeton University Press, 1960.

The Theorem of Von Neumann of section 3.3 and its original proof can be found in the above two references.

[100] Wimmer, H.K., Bezoutians of polynomial matrices and their generalized inverses, *Linear Algebra and Its Applications,* 122/123/124(1989), 475–487.

The Bezoutian of a quadruple $(F, G; U, D)$ of polynomial matrices in a variable $z$ with coefficients from a field is studied. Under certain coprime conditions it was shown that the Bezoutian has a block Hankel matrix as a $g$-inverse.

# Index

**Other titles in the Algebra, Logic and Applications series**

**Volume 8**
Multilinear Algebra
*Russell Merris*

**Volume 9**
Advances in Algebra and Model Theory
*Edited by Manfred Droste and Rüdiger Göbel*

**Volume 10**
Classifications of Abelian Groups and Pontrjagin Duality
*Peter Loth*

**Volume 11**
Models for Concurrency
*Uri Abraham*

**Volume 12**
Distributive Models and Related Topics
*Askar Tuganbaev*

**Volume 13**
Almost Completely Decomposable Groups
*Adolf Mader*

**Volume 14**
Hyperidentities and Clones
*Klaus Denecke and Shelly L. Wismath*

**Volume 15**
Introduction to Model Theory
*Philipp Rothmaler*

**Volume 16**
Ordered Algebraic Structures: Nanjing
*Edited by W. Charles Holland*

**Volume 17**
The Theory of Generalized Inverses Over Commutative Rings
*K.P.S. Bhaskara Rao*